Safe Intrusive Activities on Land Potentially Impacted by Contamination

Safe Intrusive Activities on Land Potentially Impacted by Contamination

Prepared by
British Drilling Association

Published by Emerald Publishing Limited, Floor 5, Northspring, 21-23 Wellington Street, Leeds LS1 4DL.

ICE Publishing is an imprint of Emerald Publishing Limited

Other ICE Publishing titles:
UK Specification for Ground Investigation.
The Association of Geotechnical and Geoenvironmental Specialists (AGS). ISBN 978-0-7277-6523-9
CESMM4 Revised: Civil Engineering Standard Method of Measurement
Institution of Civil Engineers. ISBN 978-0-7277-6440-9
ICE Specification for Piling and Embedded Retaining Walls, Third edition
Institution of Civil Engineers. ISBN 978-0-7277-6157-6

A catalogue record for this book is available from the British Library

ISBN 978-0-7277-3507-2

© Emerald Publishing Limited 2024

Cover photo: Ground investigation by Geotechnical Engineering Limited on land potentially impacted by contamination. Courtesy of Equipe Training Limited

Commissioning Editor: Michael Fenton
Assistant Editor: Cathy Sellars
Production Editor: Sirli Manitski

Typeset by Manila Typesetting Company
Index created by David Gaskell
Printed and bound by CPI Group (UK) Ltd, Croydon, CR0 4YY

Contents

Acknowledgements

British Drilling Association Working Groups

The British Drilling Association (BDA) extends thanks to the individuals and organisations who made up the Working Groups involved with previous versions of this document for giving their time, experience and expertise.

This document has been developed by a number of Working Groups formed by the BDA, and special thanks are extended to the following members of these Working Groups

Paul McMann	Dunelm Geotechnical & Environmental & BDA
Martyn Brocklesby	Geotechnical Engineering & BDA
Derek Smith	Coffey Geotechnics & SISG
Nick Vater	SafeLane Global & BDA
Adam Putt	Socotec
James Wilson	Fugro
Jon Christie	BAM Nuttall
Julian Lovell	Equipe Group & BDA & AGS
Darren Makin	Socotec & AGS
Roger Clark	Marlowclark & SiLC
Caroline Martin	Royal HaskoningDHV & AGS
Nigel Snedker	WSP
Ian Evans	Arcadis & SiLC

The BDA acknowledges the technical contributions made during the drafting of the document by many other colleagues of the Working Group members and members of the BDA Safety, Training and Education and Technical and Standards Sub-Committees.

Foreword

This guide has been compiled to provide health, safety and environmental information and recommendations on good practice for all those engaged with intrusive activities on land potentially impacted by contamination. These include any individuals or organisations involved in such work's specification, procurement, execution or supervision. The scope of the guidance provides information relevant to all intrusive activities, including ground investigations, piling, land drilling, earthworks and ground remediation.

Clients and designers have a duty of care to provide complete and accurate information, as early as possible, to allow contractors to develop safe systems of work to eliminate the potential impact of contamination on human health and the environment.

This document promotes safe working practices and improves awareness of health, safety and environmental matters. Land potentially affected by contamination contains particular hazards that require the employment of specialist geoenvironmental services with appropriately trained and experienced office and site personnel, suitable plant and equipment, and high levels of supervision and response.

There is no intention that this publication should replace any acts, regulations, codes or other legislative or contractual documents which have legal standing.

The Health and Safety Executive (HSE) publishes guidance to assist employers in implementing the minimum legal requirements imposed on them by acts, regulations and codes (or Approved Codes of Practice (ACoPs)). The HSE cannot publish guidance specific to any industry and, as such, industry bodies such as the British Drilling Association (BDA) create the benchmark for good practice. The guidance in this document is not a legal requirement, but the HSE advises that organisations following it should be doing enough to comply with the law. Where appropriate, the good practice recommendations contained herein should be incorporated into the contract documentation at procurement stage where there is a likelihood of intrusive activities on land potentially impacted by contamination.

This guidance assumes that the majority of intrusive activities will be carried out in accordance with the Construction (Design and Management) Regulations 2015 but, where these may not apply, to ensure good practice, it is assumed duty holders will adopt similar roles and responsibilities.

This publication should be read in its entirety to optimise all aspects of managing health and safety. Selective reading or part implementation of its recommendations will not ensure the protection of individuals to the fullest degree possible. All parties should agree to the publication as good practice.

This guidance is structured for reading as presented chronologically: firstly examining the legislative background and then detailing each stage of activity from pre-construction through to completion of site works.

British Drilling Association
ISBN 978-0-7277-3507-2
https://doi.org/10.1680/sialpic.35072.001

Safe Intrusive Activities on Land Potentially Impacted by Contamination

1. Introduction

In 1992, the British Drilling Association (BDA) published *Guidance Notes for the Safe Drilling of Landfills and Contaminated Land*. This was in response to a significant increase in drilling requirements for these situations and the absence of national guidelines covering these operations.

The Site Investigation Steering Group (SISG) subsequently adopted the BDA Guidance Notes, sponsored by the Institution of Civil Engineers. With minor modifications, they were republished in 1993 as Part 4 *Guidelines for the Safe Investigation by Drilling of Landfills and Contaminated Land*, this being the last part of the Thomas Telford publication series *Site Investigation in Construction*. Since this was published, there has been a great deal of change, such as

- increased statutory and regulatory legislation
- changes in working practices resulting from the above, particularly in procurement, to reflect health and safety requirements
- more prominence given to continuous/dynamic risk assessment
- wider understanding of what constitutes a hazard
- greater reuse of 'brownfield sites'
- an increase in awareness of the environment and environmental concerns
- increased liability and insurance issues
- technological changes in equipment and methods
- increased focus on competence and qualifications.

This publication replaces SISG Part 4 (1993) and is based on the BDA publication *Guidance for Safe Intrusive Activities on Contaminated or Potentially Contaminated Land* (published in 2008), which it also replaces.

This revision emphasises that safe working practices should be paramount to those procuring, contracting and operating on sites where intrusive activities are likely to encounter contaminated land. The Health and Safety Executive (HSE) maintains that cost constraints are not a defence in the absence of safe working practices. There is a clear requirement for organisations to recognise the need to operate a 'safe system of work' as the main priority and not select a contractor by cost criterion only.

The importance of designing health, safety and environmental matters into a project from its inception is a legal and best practice requirement. All parties should exchange information and knowledge, consult at all levels and agree – in writing – on all matters of health, safety and environment before, throughout and after site works.

This publication has retained but reviewed the BDA site categorisation colour coding system (Green, Yellow or Red) introduced in 1992. This system has been extensively adopted in the industry and helps to provide readily understandable guidance regarding the potential hazards that parties may encounter on site and the precautions to be considered. Consideration was given to introducing additional colour categories, particularly between Yellow and Red, however, the decision was taken that the existing three categories adequately indicate low-, medium- and high-risk ratings.

This revised guidance gives more emphasis to pre-construction activity before determining site categorisation. The importance of detailed desk studies, the preparation of the health and safety pre-construction information (PCI) (as required by the Construction (Design and Management) Regulations 2015 (CDM Regulations) and the Management of Health and Safety at Work Regulations 1999), preparatory works, induction, competence and risk assessment are all stressed and explained.

Activities following intrusive works, such as the handling and testing of potentially contaminated materials, are considered more prescriptively than previously.

This publication provides guidance for clients, designers, contractors, site operatives and others who may be exposed. Specific advice, such as the harmful or hazardous effects of contaminants, is not given – parties should obtain this additional information from appropriate technical references and specialists in land contamination.

These guidance notes do not relieve anyone of their responsibility to consult the various acts, regulations, standards and industry guidance that are relevant.

This publication does not include the presence of radioactive materials on a site. These are the subject of the Radioactive Substances Act 1993, the Ionising Radiations Regulations 2017 and the HSE's 2018 Approved Code of Practice (ACoP) *Working with Ionising Radiation*.

When the CDM Regulations apply to the intrusive activity, the roles and responsibilities of the client, principal designer, designer, principal contractor and contractor are well defined. However, it is recognised that a small percentage of intrusive activities may fall outside these requirements and, in these instances, this guidance assumes duty holders will adopt similar roles. Therefore, for consistency, the terms client, designer and contractor have been used throughout this document and imply the same duties as those defined in the CDM Regulations and as detailed in Section 4.0.

The application of these guidance notes has been widened to include all intrusive activities, methods and processes involved in land potentially impacted by contamination. This addition acknowledges that the original publication was directed to ground investigation only.

Regarding drilling into landfills, parties should also refer to the Environmental Services Association publication *Drilling into Landfill Waste* (ICoP 4). ICoP 4 relates explicitly to situations where a potentially explosive atmosphere is, or could be, present and is intended as guidance to assist the waste management industry in meeting the requirements of the Dangerous Substances and Explosive Atmospheres Regulations 2002 (DSEAR). In 2013, the HSE issued an ACoP (L138), which is primarily for an informed and experienced audience such as health and safety professionals. Also in 2013, the HSE produced the leaflet *Controlling Fire and Explosion Risks in the Workplace* (INDG370), which provides a short guide to DSEAR for small and medium-sized businesses.

There is a legal requirement to obtain Coal Authority permission before any work activities proceed with respect to coal seams, coal workings and mine entrances. Any activities that intersect, disturb or enter any of the Coal Authority's interests require prior written permission from the Coal Authority. The client needs to obtain a formal permit prior to commissioning any intrusive activities (see www.coal.gov.uk for details). Further guidance is provided in the 2019 Coal Authority publication *Guidance on Managing the Risk of Hazardous Gases when Drilling or Piling Near Coal*.

There is a further requirement for all planned drilling greater than 30 m deep and within 1 km of mining activities to be reported to HM Inspector of Mines in accordance with the Borehole Sites and Operations Regulations 1995.

It is the responsibility of users to ensure they are compliant with all the legislation relating to the scope of the works.

British Drilling Association
ISBN 978-0-7277-3507-2
https://doi.org/10.1680/sialpic.35072.003
Emerald Publishing Limited: All rights reserved

2. Legislation

2.1. General

The recommendations contained in this publication were compiled considering current legislation in force in the UK at the time of preparation. Variations are likely to occur over time and in other countries, and therefore reference should be made to the appropriate current national legislation.

Management of organisations and their employees must always observe health and safety legislation and regulations and consider the safety of others, including the public, who may be affected by any activity.

Sections 2.2 and 2.3 provide brief descriptions of current primary legislation (Acts) and secondary legislation (Regulations) applicable to these guidance notes. However, the list is not inclusive of all legislation that may need to be consulted. The BDA publication *Health and Safety Manual for Land Drilling: A Code of Safe Drilling Practice*, published in 2002 and republished in 2015, should also be consulted.

Practical guidance on applying specific regulations is published as Approved Codes of Practice (ACoPs). These ACoPs guide matters relating to health and safety and are approved in writing by the Health and Safety Executive (HSE) in accordance with section 16.0 of The Health and Safety at Work etc. Act 1974. An ACoP has a special legal status in that anyone subject to legal proceedings and proved in court not to have followed the relevant provisions of the ACoP can be found guilty unless they can show they have complied with the law in some other way.

This document focuses on the health, safety and environmental aspects of intrusive activities on land potentially impacted by contamination. In addition, there is a great deal of legislation related to the environment that readers should also consider. Individuals must ensure that they are aware of the current national legislation and practices, some of which are now discussed.

2.2. Acts

The Health and Safety at Work etc. Act 1974 (HASWA)

HASWA applies to all types of work activity in the workplace and imposes on organisations and self-employed persons the duty to ensure, so far as is reasonably practicable, the health, safety and welfare of their employees and others, including the public who may be affected by the work activity. HASWA also places the same responsibility on employees. It requires all organisations that employ more than five persons to prepare and revise as necessary a written safety policy and communicate its content to all employees. HASWA also permits the Secretary of State or other ministers to make regulations relating to health and safety matters. Section 2 of HASWA sets out the duty of care required by organisations concerning the health and safety of their employees while at work, and includes such things as

- the provision and maintenance of plant and systems of work that are safe and without risks to health
- ensuring that the use, handling, storage and transport of articles or substances is safe and without risk to health
- the provision of information, instructions, training and supervision for employees.

The Control of Pollution (Amendment) Act 1989

The Control of Pollution (Amendment) Act 1989 relates to construction sites and gives power to local authorities to control noise. Under the Act, they may serve notices that

- restrict the use of specific items of a contractor's plant or equipment
- restrict certain hours of working
- impose noise limits during certain periods of the day or week.

This Act also makes it an offence to cause or knowingly permit pollution of inland or coastal waters.

The Environmental Protection Act 1990

The Environmental Protection Act 1990 deals with the control of significant emissions to air, land and water from scheduled processes. All waste in the classification of 'controlled waste' must be disposed of correctly at an approved waste disposal site and accompanied by a complete description of the waste.

The Environment Act 1995

The Environment Act 1995, with the insertion of section 57 into the Environmental Protection Act 1990 as Part IIA, gives greater clarity and emphasis is given to contaminated land issues and its legal administration. Additional legislation is included in

- the Contaminated Land (England) (Amendment) Regulations 2012
- the Contaminated Land (Scotland) Regulations 2005
- the Contaminated Land (Wales) (Amendment) Regulations 2012.

2.3. Regulations

At least 29 sets of regulations may be relevant in part or whole. Some of the more important for this guide are as follows.

The Management of Health and Safety at Work Regulations 1999 (MHSWR)

The MHSWR impose a duty on an organisation to assess its entire operation from a safety perspective. The duty applies to most types of work activity and includes the requirement to

- assess the risk to the health and safety of employees and other persons who could be affected by the work activity
- institute measures shown by the risk assessment to be necessary for the management of health and safety
- appoint competent persons to carry out these activities
- set up emergency procedures
- supply employees with adequate information regarding health and safety
- ensure that all employees are suitably trained in health and safety matters
- provide welfare.

The Construction (Design and Management) Regulations 2015 (CDM Regulations)

The CDM Regulations impose duties upon all parties involved in construction work, including site clearance, exploration and investigation (but not site survey). The CDM Regulations apply to all construction activities; therefore, all types of contracts carried out in relation to civil engineering construction or building works (including any intrusive investigation or other forms of intrusive activity) would be subject to the regulations. Compliance with the CDM Regulations is therefore mandatory for intrusive activities following this definition. The CDM Regulations aim to integrate health and safety into the management of a project and to encourage everyone to work together to

- improve the planning and management of projects from the very start
- identify risks early on so that they can be eliminated or reduced at the design or planning stage and the remaining risks can be properly managed
- target effort where it can do the best in terms of health and safety
- discourage bureaucracy

The Construction (Head Protection) Regulations 1989

The Construction (Head Protection) Regulations 1989 require organisations to provide, maintain and replace, as necessary, suitable head protection for employees and other persons working or entering the working area for which they have responsibility. They must ensure that head protection is worn unless there is no risk of head injury other than from falling over.

The Personal Protective Equipment at Work Regulations 1992

The Personal Protective Equipment at Work Regulations 1992 refer to personal protective equipment (PPE). This includes waterproof clothing, gloves, safety footwear, high-visibility jackets and waist-coats, eye protection, respirators, safety harnesses and the like. Every organisation is required to provide suitable PPE to every employee who is exposed to risk at work.

The Control of Substances Hazardous to Health (Amendment) Regulations 2004 (COSHH)

These regulations provide a framework for controlling substances at work that may be hazardous to health, for example toxic substances and substances that cause irritation or are corrosive. Organisations must assess the risks from the hazardous substance and decide on control measures to prevent harmful exposure.

The Reporting of Injuries, Diseases and Dangerous Occurrence Regulations 2013 (RIDDOR)

These regulations are concerned with reporting incidents that occur during or because of work activity. Responsibility for reporting the incident lies with the 'responsible person' as defined in the regulations or the person in charge. The HSE also provides a brief guide to RIDDOR in INDG453. Amendments specific to the quarrying industry are provided in the 1999 HSE *Guide to the Reporting of Injuries, Diseases and Dangerous Occurrences Regulations 1995* (L73).

The Health and Safety (First Aid) Regulations 1981

These regulations set out organisations' duties for providing first aid at the workplace.

The Workplace (Health, Safety and Welfare) Regulations 1992

These regulations set out the minimum legal requirements for providing workplace health, safety and welfare standards, which include providing welfare facilities, lighting, heating and ventilation, access and pedestrian and traffic routes, drinking water supplies and protection from falls or items falling from height.

The Control of Lead at Work Regulations 2002

These regulations define occupational exposure limits, risk assessment procedures, control measures, care and decontamination of PPE and the organisation's duty to prepare emergency procedures.

The Control of Asbestos Regulations 2012

These emphasise risk assessment, control measures and emergency procedures.

British Drilling Association
ISBN 978-0-7277-3507-2
https://doi.org/10.1680/sialpic.35072.006

3. Competence, training and qualifications

3.1. General

Only competent personnel should be used to design, manage and supervise intrusive activities in contaminated or potentially contaminated land. They should have sufficient and relevant experience, be appropriately qualified and trained to carry out the work safely and capable of recognising new or unexpected risks as they arise.

Preferentially, competent personnel are also recommended in relation to all activities where practicable. However, it is permissible that trainees or less experienced personnel may be used in low-risk situations under competent supervision. The competent person represents the responsible person in these circumstances.

Competent personnel are expected to know how to deal with known, potential or exposed hazards, thus allowing the management and control of hazards, together with the minimisation of risk to any person, whether on site or potentially affected by site operations.

3.2. Competence

Competence is a combination of

- training
- knowledge
- experience
- skill
- attitude.

All organisations must ensure that their staff are competent to carry out their roles safely, particularly with providing health and safety assistance, as required by HASWA.

Designers and contractors should keep records of the competence of all persons, including dates and contents of training, qualifications, certification and experience. This information should be provided to clients upon request. Records may also need to be provided in the event of accidents or incidents.

Where specific competencies are required for the intrusive work, these should be identified by the specifier or designer in the specification. If the AGS 2022 publication *UK Specification for Ground Investigation* is adopted, these details should be provided within the schedules.

3.3. Qualifications

Land drilling lead drillers need to hold or be registered and working towards an appropriate National Vocational Qualification (NVQ) for the type of machine being used and their work activity. This should either be a Level 2 Diploma in Drilling Operations – Lead Driller or a Level 3 Diploma in Advanced Land Drilling. Drilling support operatives shall hold or be registered and working towards a Level 2 Diploma in Drilling Operations – Drilling Support Operative.

Land drilling operatives should hold a BDA Audit Card as proof of current competence. This card is renewable every 12 months after on-site auditing by the BDA. It displays the job category (lead driller or drilling support operative) and, for lead drillers, the drilling categories for which they have been audited. BDA Audit Cards are only issued to NVQ land drilling qualified personnel with a Construction Skills Certification Scheme (CSCS) Blue Skilled Worker or Mineral Products Qualifications Council (MPQC) competency card in land drilling.

The BDA Audit Card should mirror the NVQ certificate and CSCS/MPQC card, detailing the relevant pathway, endorsement(s) and category(s) taken from the following

- pathway – lead driller or drilling support operative
- endorsements (for lead drillers only) – ground investigation, grouting, soil nailing and anchoring, water wells, landfill, geothermal, dewatering, extractives
- category (for lead drillers only) – cable percussion, rotary, sonic, dynamic sampling.

Piling rig operators should hold or be registered and working towards a Level 2 Diploma in Piling Operations – Piling Operative and piling attendants should hold or be registered and working towards a Level 2 Diploma in Piling Operations – Piling Attendant. In addition, piling rig operators should hold a valid and current Construction Plant Competence Scheme (CPCS) card relevant to the size of piling rig being operated.

No lead driller or piling rig operator should operate the machine without the correct competence, unless supervised by a competent person.

Every site worker should also hold either a CSCS skills, MPQC competency or CPCS skills card for which they have passed a Construction Industry Training Board (CITB) approved health, safety and environmental test, satisfying the requirements of their attained card.

The validity of a CSCS, CPCS or MPQC card can be checked with the issuing body or through smart card validation, where available. The validity of a BDA Audit Card can be checked by viewing the list of audited drillers on the BDA website or by clicking on the BDA Audit smart e-card held by the operative.

3.4. Training

All personnel working on site are required to have received relevant instruction and training from their employer. Additional task-specific or hazard-specific training may be required depending upon the nature and complexity of the work on the land potentially impacted by contamination.

Supervisors should attend a training course relevant to the work activity they are supervising, such as the IOSH (Institution of Occupational Safety and Health) Safe Supervision of Geotechnical Sites or, for construction-related activities, the CITB Site Supervisor's Safety Training Scheme. The supervisor or site manager should conduct a site safety induction (see Section 15.0) to explain the site-specific method statement and risk assessment based on the significant hazards identified from the desk study. Desk studies are discussed in greater detail in Section 5.0.

All site personnel should receive training in the symptoms of exposure to the principal contaminants, vapours and gases, and instruction on where and how to obtain treatment.

All personnel working on a landfill site should receive advice on diseases associated with such environments (such as leptospirosis (Weil's disease), tetanus etc.), as well as other ecological hazards. Similarly, they should receive advice on the health hazards associated with contaminated land.

Where asbestos is suspected or likely, site personnel should receive asbestos awareness training as a minimum or, depending upon the site-specific conditions and/or procedures, particularly when sampling is required, should have asbestos non-licenced training. More details can be found in the Control of Asbestos Regulations 2012, the 2021 AGS guidance on the *Assessment and Control of Asbestos Risk in Soil* and the 2016 CL:AIRE document *CAR-Soil*. Courses should relate to asbestos in soil, for example Equipe's Managing Asbestos Risk in Ground Investigation, CL:AIRE's Managing and Working with Soil and Construction & Demolition Materials Affected by Asbestos or UKATA's Asbestos Awareness in Soil and Made Ground.

These guidance notes recommend that all site operatives receive, as a minimum, Emergency First Aid at Work training, a one-day course run by approved organisations such as the British Red Cross, St Andrew's First Aid and St John Ambulance. All BDA-audited lead drillers and drilling support operatives are required to have attended the one-day Emergency First Aid at Work course.

Where flammable or explosive gases are anticipated or suspected, site personnel should receive training on how to use gas-detecting systems and what actions they must take should gas levels reach an action limit or the alarm sounds.

Site personnel should receive training in using respiratory equipment and other specialised PPE that they may require on contaminated sites. They should have also been face fitted for the specific respiratory protective equipment (RPE) to be used.

Prior to the set up and use of a decontamination unit, all site personnel should attend a familiarisation briefing and be made aware of the arrangement, set up and operation of the decontamination unit. This should include, as a minimum, the layout of the unit, rules for correct use and operation of the extraction system.

British Drilling Association
ISBN 978-0-7277-3507-2
https://doi.org/10.1680/sialpic.35072.009

4. Managing health, safety and the environment

Directors and senior management have defined duties under the Health and Safety at Work etc. Act 1974 and a significant role to play in the management of all health, safety and environmental matters involved with intrusive activities on land potentially impacted by contamination. They must have a commitment to health and safety, including initiating and participating in safety programmes, providing instructions and training personnel as well as ensuring safe machinery, tools and working conditions.

All employees also have a duty to take reasonable care of their own safety and the safety of any others who may be affected by their acts or omissions and also to co-operate with their organisation and others to enable them to comply with their duties. This also prohibits the intentional or reckless interference with, or misuse of, anything provided in the interests of health, safety and welfare.

On every site, an individual must be designated to record any accidents or illnesses. Personnel should report feelings of nausea, irritation to the throat, nose, eyes, headaches and so on to this designated person. The designated person must have at least basic first aid training and be permanently on site during intrusive works or be able to hand over the responsibility to another suitably qualified individual. Any records made must be handed to the employer at the end of the intrusive work.

On any project, consideration must be given to appropriate welfare facilities for site operatives under HSE guidelines.

The flow diagram in Appendix D has been prepared to identify the critical health and safety management processes involved in commissioning, designing and safely undertaking a project.

All personnel should be empowered to stop work on any project should any unexpected hazards, including types or levels of contamination, be encountered. Staff should also contribute ideas to improve working procedures and practices as stated in the Health and Safety (Consultation with Employees) Regulations 1996.

If a hazard assessment or risk register has been carried out, the client must make a copy available to the contractor.

Where the CDM Regulations apply, the requirements for good management of health, safety and environment are well defined, along with what each duty holder must or should do to comply with the law to ensure projects are carried out in a way that secures health and safety. The roles and responsibilities of the main duty holders can be summarised as follows.

Clients are organisations or individuals for whom a construction project is carried out. They make suitable arrangements for managing a project, including making sure that

- other duty holders are appointed
- sufficient time and resources are allocated
- relevant information is prepared and provided to other duty holders
- the principal designer and principal contractor carry out their duties
- welfare facilities are provided.

Principal designers are designers appointed by the client in projects involving more than one contractor. They can be an organisation or an individual with sufficient knowledge, experience and ability to carry out the role. They must plan, manage, monitor and coordinate health and safety in the pre-construction phase of a project, including

- identifying, eliminating or controlling foreseeable risks
- ensuring designers carry out their duties

- preparing and providing relevant information to other duty holders
- providing relevant information to the principal contractor to help them plan, manage, monitor and coordinate health and safety in the construction phase.

Designers are those who, as part of a business, prepare or modify designs for a building, product or system relating to construction work. They provide information to other members of the project team to help them fulfil their duties when preparing or modifying designs in order to eliminate, reduce or control foreseeable risks that may arise during construction and during the maintenance and use of a building once built.

Principal contractors are contractors appointed by the client to coordinate the construction phase of a project where it involves more than one contractor. They must plan, manage, monitor and coordinate health and safety in the construction phase of a project, including

- liaising with the client and principal designer
- preparing the construction phase plan
- organising co-operation between contractors and coordinating their work
- ensuring suitable site inductions are provided
- ensuring reasonable steps are taken to prevent unauthorised access
- ensuring workers are consulted and engaged in securing their health and safety
- ensuring welfare facilities are provided.

Contractors are those who do the actual construction work. They can be individuals or a company. They must plan, manage and monitor construction work under their control so that it is carried out without risks to health and safety. For projects involving more than one contractor, they must coordinate their activities with others in the project team – in particular, they must comply with directions given to them by the principal designer or principal contractor. For single-contractor projects, a construction phase plan must be prepared.

The above does not relieve the contractor of legal duties or responsibilities concerning the health and safety of their employees. If the client or the designer cannot provide the necessary information, then the contractor should take reasonable steps to do so and be reimbursed accordingly.

British Drilling Association
ISBN 978-0-7277-3507-2
https://doi.org/10.1680/sialpic.35072.011

5. Desk studies

A comprehensive desk study is recommended to provide relevant information that can be used in the preparation of the pre-construction information (PCI) under the CDM Regulations. The desk study should be completed in accordance with the Environment Agency's 2021 publication *Land Contamination Risk Management*, BS 5930:2015+A1:2020, BS 10175:2011+A2:2017 and guidance documents published by the Association of Geotechnical and Geoenvironmental Specialists (AGS), the Building Research Establishment (BRE) and the Construction Industry Research and Information Association (CIRIA) (see Section 20.0).

The desk study information is used to prepare a robust PCI document that must provide essential health, safety and welfare information and highlight potential environmental issues. It must be carried out at the earliest opportunity and always before commencement of the works. The various publications referred to above describe the components of a desk study. The following, which is not an exhaustive list, should be considered in the desk study

- current site usage and historical land use information
- records of contact with the local authority's contaminated Land Officer/Environmental Health Officer
- reference to bespoke guidance of the relevant planning authorities (as available)
- a review of unexploded ordnance (UXO), mining, radon and geological risks
- assessment of the site's environmental information, including but not limited to geology, waste facilities, water abstractions, pollution incidents, permits, environmental ecologically sensitive areas, hydrology and hydrogeological sensitivity
- reference to web-based resource information, including but not limited to the British Geological Survey, the Coal Authority, the Environment Agency/Natural Resource Wales/Scottish Environmental Protection Agency and Magic Map
- before the investigation commences, a walkover will complement understanding of the site setting, access, restrictions and its boundary conditions (e.g. adjoining land use)
- a review of third-party relevant reports (if available)
- anecdotal evidence (e.g. information from those previously involved with the site), although judgements may need to be made on the likely validity of this information
- details of the site under consideration as well as the immediately surrounding land that potentially has similar contamination or underlying geological risks
- a geotechnical risk register and a contaminated land preliminary risk assessment.

A competent person who is familiar with the process and has knowledge of the sources from which the information was obtained should carry out the desk study.

Where preliminary works include minor intrusive activities and a comprehensive desk study is not available, the client should be made aware that critical information may be missing and their legal obligations may not be fulfilled. In any event, sufficient information must be provided to the contractor to allow full evaluation of the hazards and development of suitable risk assessment in order to plan the works.

British Drilling Association
ISBN 978-0-7277-3507-2
https://doi.org/10.1680/sialpic.35072.012

6. Risk assessment

6.1. General

Intrusive activities on land potentially impacted by contamination can present risks to

- site staff involved in the work
- third parties in the vicinity of the work
- the public
- the environment.

The principal risks to human health during work on land potentially impacted by contamination include

- inhalation of toxic gases and vapours
- inhalation of fugitive dust containing contaminants
- dermal (skin) contact with contaminated material
- ingestion of contaminated material.

Other risks are associated with intrusive works, and parties should reference the 2015 BDA publication *Health and Safety Manual for Land Drilling: A Code of Safe Drilling Practice*. Such risks would include – but are not limited to – overhead and underground services, UXO, ground gas, vapours and landfill- and mining-related gases.

The principal risks to the general environment could include

- artesian water
- runoff of contaminated drill flush to controlled waters
- the creation of pathways to allow contamination to enter uncontaminated ground or groundwater
- the production of contaminated arisings and wash-water from site operations, with a consequent requirement for controlled disposal
- discharges of effluent to surface waters and ground
- the introduction of hazardous substances by way of drilling fluids, lubricants and concrete.

It is vital that any works on land potentially impacted by contamination are planned adequately to identify all the risks associated with the hazards assessed and a method of working designed to manage them.

The risk management process should be carried out before commencing any intrusive activities works, including

- *hazard identification* – identifying the hazards that may be associated with the site
- *hazard assessment* – assessing the hazard associated with the site (i.e. the potential exposure and impact to site workers, third parties or the environment, assuming plausible pathways exist or will be created)
- *risk estimation* – estimating the likelihood that an adverse effect will result from exposure to the hazard and the impact
- *risk evaluation* – evaluating the significance of estimated risks, taking into account available guidelines, standards and the uncertainties associated with the assessment
- *control measures* – implementing a method of working to mitigate and manage risks.

This process is commonly referred to as a 'risk assessment' and is further explained in the HSE's *Risk Assessment: The Key to Good Health and Safety*.

Risk assessment is a statutory requirement under the management of health and safety regulations. PCI (including health and safety information) must be prepared, highlighting the principal hazards and the identified risks associated with those hazards at the design stage.

To ensure that appropriate procedures are in place, the site should be classified at the pre-tender stage regarding risk to workers from contamination. An example of good practice would be to use the established BDA site categorisation to assess and manage the risks associated with the site accordingly (see Section 8.0). The decision-making should be recorded and made available to those at risk.

Legally, the contractor remains responsible for developing and maintaining a construction phase plan (or equivalent) for the duration of the construction works. As part of this plan, contractors will need to carry out a risk assessment, which will often be specific to their particular method of working and internal management systems. The CDM Regulations define the construction phase plan as a document recording the health and safety arrangements, site rules and any special measures for carrying out the construction work.

Notwithstanding the above, the contractor has legal responsibility for the health and safety of its employees. If the client or designer cannot supply the necessary information, then the contractor should carry out independent research to determine the likely risks posed to the workforce associated with any hazards that might be present.

Small contracts can present risks to personnel that are the same or greater than those associated with larger contracts.

6.2. Hazard identification

A 'hazard' is the property of a substance or situation that has the potential to cause harm. Hazard identification involves the initial identification of potential contaminants, pathways and receptors. Hazard identification and hazard assessment do not consider the likelihood of harm – this is determined by risk estimation and risk evaluation. Research must be conducted to identify potential health and safety hazards.

A site reconnaissance visit may reveal the land's current or recent historical use. Specific historical uses may suggest potential contamination of the ground and/or groundwater beneath the site. It should be recognised that industrial processes carried out in the past may have had little or no regard for the environment compared to today. Surface or underground tanks, chemical and waste storage areas, unusual odours and surface features (e.g. polluted watercourses, ground discolouration, ground staining, oily sheens on standing water, signs of apparent damage or distress to plants or the complete absence of vegetation) can all indicate the presence of contamination.

During site reconnaissance, visiting the local studies section of the region's library or sourcing maps from the internet can also be helpful. Copies of superseded County Series and Ordnance Survey maps and old trade directories can provide further clues to the historical use of a site. Parties may also source information from commercial companies specialising in holding historical records.

Once a historical, recent or current use has been determined, the Department of Environment (DoE) Industry Profiles (available on the CL:AIRE Water and Land Library) are a useful resource. Appendix B provides a summary table of past land uses and potential BDA site categorisation, but further guidance should be sought to define the potential contaminants associated with those land uses.

The DoE Industry Profiles do not include emerging contaminants that are synthetic (man-made) or naturally occurring contaminants that are not routinely being monitored in the environment but either already exist in the environment or are very likely to migrate into the environment at some point in the future. Examples include hormones, plastic components, flame retardants, surfactants, fragrances, pesticides and pharmaceuticals. Consideration should not only be given to the source chemicals but also to degraded and residual products that can arise due to natural breakdown of the source chemicals.

Recent publicity has been given to perfluorinated and polyfluorinated alkyl substances (PFAS), which are a group of substances that include perfluorooctane sulfonate (PFOS) and perfluorooctane acid (PFOA). They have been extensively used for the past 30 years in food packaging, non-stick surfaces, fire-suppression foams, waterproof clothing, wire casing and resistant tubing. They are very hard to break down, resulting in very little biodegradation in the environment, and only change concentration by dilution. They are linked to several forms of cancer and effects on the liver and the gastrointestinal system. Other examples of emerging contaminants are phthalates (with a wide range of possible health effects), micro silicones (some are persistent, bioaccumulative and toxic) and chemicals used to make explosives, but there are many more at present that will likely expand in the future due to increasing knowledge of toxicity, bioaccumulation and persistence of chemicals in the environment. The possibility of emerging contaminants being present should be taken into consideration in researching the circumstances of the site and when referring to Appendix B.

There may be potential hazards outside the site boundaries, for example the flow of pollutants along groundwater hydraulic gradients determining groundwater flow and as a result of aerial distribution, which could impact intrusive works. The research and reconnaissance should therefore also look at adjacent land uses.

Contact could be made with the local authority as part of the desk-based research. Under the Contaminated Land Regulations 2012 (for England, Scotland and Wales), the local authority is responsible for identifying potentially contaminated land within its area. Therefore, contact with the local authority might prove useful in determining potential past contaminative usage for a particular site. Similarly, it may be useful to contact the local office of the Environment Agency (EA), Scottish Environmental Protection Agency (SEPA) or Natural Resources Wales (NRW).

Access to the findings of previous ground investigations, especially where chemical analyses have also been conducted, would help reduce uncertainty regarding potential contaminants. The local authority and EA/SEPA/NRW may also be aware of any pollution incidents and/or remediation schemes that have been carried out at a site.

6.3. Hazard assessment

The hazard assessment phase of the risk assessment process considers plausible pollutant linkages and determines the potential for human health and environmental risks associated with the activities proposed.

The type of pollutant linkages or pathways by which the hazard can encounter receptors, be they human or environmental, will depend upon the characteristics of the hazard and the method of working proposed. Likely pollution pathways for site workers and members of the public include

- ingestion
- inhalation of air-borne vapours or particulates (dust)
- dermal absorption following skin contact.

Environmental pathways include the release of contaminated liquids to surface waters (e.g. by way of drill flush) and the formation of preferential pathways for contaminated water to migrate to underlying aquifers through the action of intrusive processes.

6.4. Risk estimation

A 'risk' is the probability or frequency of occurrence of a defined hazard combined with the magnitude of the consequences of that occurrence.

Risk estimation is concerned with evaluating the likelihood that contamination on a site will lead to an adverse effect on the receptor. For example, if the probability of contamination on a site is low and the consequences to the site worker encountering the level of contamination expected are not severe, then the risk would be assessed as 'low'. Conversely, if there is a high probability of contamination in the ground and the contamination involves such materials that the human health consequences of encountering those contaminants are severe, the risk is assessed as 'high'.

6.5. Risk evaluation

A suitably experienced and competent person must carry out the risk evaluation. The risk assessment process considers whether the identified risks are acceptable or require mitigation to manage them appropriately.

The results of the risk assessment should be used to determine

- appropriate action to ensure that the level of exposure to the potential hazard of all persons and the environment is kept as low as reasonably practicable (ALARP)
- the need to designate areas as controlled or supervised and to specify local site rules
- the training needs of supervisory and non-supervisory personnel
- the type of any PPE and/or RPE that may be required
- emergency procedures to be implemented in the event of exposure to a hazard.

British Drilling Association
ISBN 978-0-7277-3507-2
https://doi.org/10.1680/sialpic.35072.016

7. Intrusive activities into waste on landfill sites

The Environment Agency (EA) has provided guidance for intrusive activities specified into waste on permitted landfill sites.

This guidance provides details for the intrusive activity, which typically comprises rotary drilling, the competence of the contractor, job roles and individual competence requirements, supervisory requirements, reporting and details of the intrusive and associated activities. The planning of such works should include consideration of the intrusive activity, including

- the environmental risks of penetration of the base and sidewalls, including if a containment system is not present
- the accuracy of construction records and survey data
- drilling methods, level of control and accuracy
- the heterogeneous and unpredictable nature of waste
- saturation by leachate towards the base of the cell
- likely perched leachate levels
- the contamination potential of wastes and leachate from the borehole
- landfill gas and odour emissions
- remedial actions in the event of damage to lining systems or other infrastructure.

A plan (normally a Construction Quality Assurance plan) must be provided to the EA, outlining the work proposed. The work must not begin until the EA has confirmed that it is satisfied with the proposals. This plan must provide details of the personnel involved in the works and their level of competence. It must also include the following details about the work activity

- the frequency, accuracy and method of surveying before and throughout the intrusive activity
- the person responsible for surveying
- use of a zonal drilling approach for all holes drilled into the landfill
- demonstration that changes in levels of the cell base have been considered, particularly when drilling into side slopes or over inter-cell bunds
- a description of the method of backfilling or grouting abandoned holes
- an action plan and remediation strategy in case the containment system or base or sidewalls are penetrated.

Further details can be found at https://www.gov.uk/guidance/landfill-operators-environmental-permits/design-and-build-your-landfill-site.

British Drilling Association
ISBN 978-0-7277-3507-2
https://doi.org/10.1680/sialpic.35072.017

8. BDA site categorisation

Every site potentially affected by land contamination should be categorised so that the associated degree of hazard can be recognised. This categorisation applies in respect of known and suspected contamination.

Appendix A presents a colour coding system for BDA site categorisation (Green, Yellow and Red). This system was established in 1993 and has been widely adopted by the industry for classifying land potentially affected by contamination. Several steps need to be followed to make use of the BDA site categorisation. While these steps are discussed separately within this publication, the following paragraphs briefly describe the actions that must take place to ensure best use of the system.

A well-structured, accurate and informative desk study, including the outcome of the following risk assessment processes, is crucial to determining the appropriate site categorisation. This guide recommends that provision is made for a site walkover survey. This step will assist in contextualising the site circumstances (within its broader environment) and yield a better understanding of historical and current uses. It may also highlight issues unlikely to be established from documentation resources alone. A risk assessment should be carried out (see Section 6.0) in terms of the likely contaminants, including gas and vapours, that may be present both on the site and the land surrounding the site. After which, a BDA site categorisation should be assigned. If there is doubt as to which categorisation should be assigned after carrying out the desk study and reviewing the previous usage, the higher of the two categorisations under consideration should be taken.

Once the site categorisation has been determined, this should allow the designer and contractor to develop and document safe systems of work.

In the case that further information becomes available that warrants a change of categorisation, work should be suspended and a reassessment must be carried out. Any subsequent changes should be implemented immediately and briefed to all site personnel. This includes information from the intrusive activity itself.

Works should only recommence when it is safe to do so and additional PPE and/or safety equipment is available.

British Drilling Association
ISBN 978-0-7277-3507-2
https://doi.org/10.1680/sialpic.35072.018

9. Project specification

The specification should clearly define all the objectives for the project. In particular, the specification documents must cover the intrusive element of the works to enable the contractor to select appropriate plant and equipment, determine operational methods, choose satisfactory materials and, most importantly, provide appropriate protective measures for the workforce.

The client needs to employ an advisor with significant experience with contamination issues at an early stage to help advise on all aspects of the work. This individual could be a representative of the designer or one of the contractor's employees. This advisor must be involved in preparing the project risk assessments, with information taken from the project health and safety file and the PCI. The survey information should also be included in the health and safety file.

If the intrusive works are solely to assess contamination issues, this should be clearly stated. Any specific reasons for undertaking separate works should be noted, for example if there is a requirement for the level of contamination to be quantified before implementing a sampling regime.

Individual contractors may use different intrusive techniques. If the nature of the contamination requires a particular technique, this should be specified.

Where samples are obtained for geoenvironmental testing, the contract should specify the likely range of contaminants for which samples are to be analysed. This specification may have a bearing on the choice of the sample container, detection limits, storage times for samples kept on site, transport arrangements and sample handling.

Contaminated sites with a noted in situ soil asbestos risk (i.e. asbestos is likely in the near-surface soil sequence) need to be pre-screened before any further testing is undertaken to mitigate the associated risks. When sending samples from a site potentially impacted by contamination for geotechnical laboratory testing, the laboratory must be informed of the potential contaminants and related risks. It is recommended to highlight samples indicated on site as potentially contaminated against those where contamination is unlikely (i.e. made ground to the natural sequence). The use of colour-coded labels would be appropriate to raise awareness in those processing samples at a later point in the investigation.

Certain aspects must be specified and discussed before the commencement of work. These include site access, site-specific safety requirements, the likely nature of materials to be encountered, final hole depth and diameter, sampling and in situ testing requirements, potential problems associated with cross-contamination and the type of installations.

Schedules to the contract document (e.g. in the AGS *UK Specification for Ground Investigation*) should specify any requirement for the contractor to demonstrate, using method statements, a clear understanding of and a capability for dealing with specific health, safety, environment and welfare hazards and risks.

The project specification should include any information on risks that might affect the contractor's personnel on a site due to ongoing site processes.

The health and safety of site personnel is likely to be compromised if, in the interests of reducing costs, the project specifies an inappropriate low-cost method of intrusive operations. Intrusive techniques must be safe and appropriate to the known or likely site conditions. Low-cost techniques should not be used if deemed to be unsafe.

To prevent such occurrences, the client and designer should satisfy themselves that the specification is appropriate to the project's overall aims. This process may require consultation with the contractor regarding specialist advice on the suitable techniques for different objectives.

British Drilling Association
ISBN 978-0-7277-3507-2
https://doi.org/10.1680/sialpic.35072.019

10. Contractual requirements

Any contract is a formalised relationship between key parties or stakeholders vested in the project. In this regard, for all intrusive activities, there is clear commercial benefit in identifying the key parties at an early stage, typically within the tender information, and this should include the arrangements that exist between them.

Contractual arrangements and terminologies will vary depending upon the chosen conditions of contract to be entered into. These defined parties should not be confused with the defined roles and responsibilities that fall under the CDM Regulations, and care needs to be taken when drafting contract conditions.

Each form of contract will have named parties, including the entity requiring the work (the client) and the company taking the lead role in delivering the work (the contractor). The contractor may use other contractors, often termed subcontractors, to complete the work but, ultimately, the contractor is responsible for the whole of the intrusive activity phase and should be formally appointed. Other parties (e.g. the designer of the works) will often be the originator of the scope of works and technical requirements, and they may be appointed as part of the contract or separate to it depending upon the form used. There are other forms of appointment and the relationship between parties may vary. Nevertheless, the roles and responsibilities of each party under the contract should be clearly defined.

The Infrastructure Conditions of Contract for Ground Investigation (ICC) and New Engineering Contract (NEC)/Engineering and Construction Conditions of Contract (ECC) represent two of a number of specimen contractual documents that can be applied to intrusive activities. Contractual documents like the NEC/ECC and ICC set out the terms, conditions and agreements, the responsibilities of the various parties involved, as well as a process for dispute resolution. The intention is to ensure agreement between the parties on certain fundamental aspects of the work being procured. These include, but are not limited to

- legal responsibilities
- insurance
- work specification
- scope
- workmanship
- materials
- programme
- safety.

The specification is commonly referenced within the contractual framework. Where ground investigations form the intrusive works, it is recommended that the AGS *UK Specification for Ground Investigation* is used for this purpose. This specification provides much of the supporting information substance that can be used by a contractor in determining the best way to address the required scope of work. It also contributes to the PCI, in accordance with the CDM Regulations. PCI is normally provided as a standalone document with the expected conditions applicable to the site used in a rational risk assessment process. In addition to the specification providing details on how the investigation should be undertaken, it also includes a standard itemised costing structure, where quantities and requirements can be assessed by the contractor in order to provide rates for each aspect of the operation. Key information given in the specification schedule generally comprises

- the name of the contract, the investigation supervisor and a description
- main works and purpose of the contract
- scope of the work
- geology and ground conditions, schedule of drawings and documents

- general requirements (quality management system, supervision and management, traffic management, suspected mine workings, archaeological remains, protected species, reinstatement etc.)
- work activity (e.g. compound requirements, security, site preparation, hole/pile locations, hole/pile type, scheduled depths, reinstatement).

In terms of safety there is a similar relationship that represents a key component of all intrusive activities, as alluded to earlier, that must be compliant with the CDM Regulations. The CDM Regulations are about focusing attention on effective planning and management of construction projects, from design concept onwards. The aim is for health and safety considerations to be treated as a normal part of a project's development, not an afterthought or bolt-on extra. Inevitably there is some overlap between information, either with respect to information submitted to accompany the tender or contract, or the details required to address the objectives of the CDM Regulations. The objectives of the CDM Regulations are

- sensible planning of the work so that the risks involved are managed from start to finish
- having the right people for the right job at the right time
- the co-operation and coordination of work with others
- access to the right information about risks and how those risks are to be managed
- effective communication of information to those who need to know
- consultation and engagement with workers about the risks and how they are to be managed.

Within this relationship there are a number of important designations, known as duty holders, each playing their defined role in order to engender safe systems of work and the minimisation of risk. Duty holders include clients, principal designers and principal contractors, although other duty holders (designers, contractors and workers) are also defined and should be clearly established in advance of carrying out the intrusive activities. A ground investigation or piling contractor within this safety framework will generally either be appointed as a principal contractor or contractor depending on the situation. Where the contractor is appointed as a principal contractor by the client in a commercial scenario, they have a broad remit of responsibilities with regard to health, safety and welfare of all those involved in the project. Accordingly, the PCI information available during the pre-contract phase of the work must be of a commensurate standard to the size and complexity of the project being considered. Furthermore, in the spirit of the CDM Regulations and the stated aim of making health and safety an integral part of a project's development, ideally, the PCI should always be consulted as early as possible for any planned works.

The documentation must contain all relevant information that will enable the contractor to better assess the health and safety risks at tender stage and must also be sufficient to competently fulfil appointments under the CDM Regulations in the event of award of the contract. Failure to provide this information is clearly not compatible with good practice, would be considered negligent by fellow professionals and fail to meet legal obligations. The following list of relevant information is deemed to have importance to the contractor, but this should not be considered exhaustive; the actual information is largely dictated by the project setting, the complexity of the operations, technical requirements and project scale

- pre-construction information (PCI)
- site desk study (including walkover information)
- underground and above-ground utility and services information (e.g. PAS 128:2022 underground utility detection survey)
- historical maps and plans
- information on sources of contamination (known or inferred)
- contamination test results
- geological information
- hydrogeological information
- UXO risk information
- mining/extractive processes information.

For large projects, this may constitute a considerable amount of transferrable data and in this instance a summary would be a preferred option to condense any salient findings. Should the originator of the information (typically the designer) take this route, then the summary should also be made available to the contractor and, where required, access to the original documents permitted for a second opinion. All such information received by the contractor should be forwarded at tender stage to any potential subcontractors who may be involved with the project. This also applies to any subcontractors that are engaged subsequent to the contract award.

The absence of relevant details concerning the site (as outlined above) could severely impair the contractor's ability to undertake detailed risk assessments for all activities required within the scope of works. Furthermore, from the perspective of the contractor being appointed to the role of principal contractor, such impairment would affect the adequacy of the construction phase plan prepared in accordance with the CDM Regulations. In the absence of information deemed crucial to the health, safety and welfare of the project, the contract should make provision for the contractor to obtain such information within the pricing structure. Any information supplied with the contract documents for tendering purposes should be clearly referenced with regard to date and origin. The information should also be concise and specific to the application. Maps and plans showing the locations of previous exploratory holes or sampling points will help the contractor assess the relevance of the data in relation to the current project. Any information/data that is known to be incomplete or potentially unreliable should be identified as such in the PCI or contractual documents in which they are referenced.

British Drilling Association
ISBN 978-0-7277-3507-2
https://doi.org/10.1680/sialpic.35072.022

11. Insurance

All organisations undertaking intrusive activities on land potentially impacted by contamination must hold valid and appropriate levels of insurance cover. Provision of evidence of this cover should be a requirement for submitting a satisfactory bid or offer for carrying out the work. The same applies to any subcontractors.

The insurances that an organisation should hold include employer's liability (this is a mandatory legal requirement) and public liability (often public and products). In addition, contractors' all risks, professional indemnity, hired in plant and motor vehicle insurance may also be required.

In general, insurers have indicated that they consider any work on landfills or contaminated land to present them with an increase in risk. The non-disclosure of such a material fact may invalidate the cover. Therefore, this guide recommends that ground investigation contractors and consultants notify their insurers that they are involved in such work.

British Drilling Association
ISBN 978-0-7277-3507-2
https://doi.org/10.1680/sialpic.35072.023

12. Health and safety planning

Health and safety planning should be carried out before intrusive activities and should be fully documented in a site-specific format. Where the contractor undertakes a principal contractor role under the CDM Regulations, a construction phase plan is required to fulfil regulatory requirements. In other circumstances, where this is not the case, safety documentation remains mandatory and equivalent documentation should be prepared. These equivalent documents are often very similar in content to a construction phase plan, such as the site-specific risk assessments and method statements (RAMS). The planning will form a basis for the management and control of safety when working on site.

The CDM Regulations define construction work as the carrying out of any building, civil engineering or engineering construction work and this includes, among others, 'the preparation for an intended structure, including … exploration, investigation … and excavation'. This definition includes intrusive activities such as ground investigations, piling, ground remediation and earthworks.

Regulation 3 of the CDM Regulations applies to all construction work regardless of the size, number of people and nature of the project, and covers general management duties. It includes a requirement that clients promptly provide PCI to every contractor who 'may be appointed by the Client'.

The PCI comprises relevant information that contractors are likely to need to plan and manage their work. Clients must provide designers and contractors who may be bidding for the job (or who they intend to engage) with project-specific health and safety information needed to identify hazards. This information should comprise

- a description of the project
- the client's considerations and management requirements
- environmental restrictions and existing on-site risks
- significant design and construction hazards
- the health and safety file.

The level of detail in this information should be proportionate to the risks in the project.

Parties should consult Appendix 2 of the HSE guidance on the 2015 CDM Regulations (L153) for topics to consider when drawing up the PCI.

The PCI for intrusive activities works may include, but is not limited to

- BDA site categorisation (e.g. Green, Yellow or Red categorisation)
- the presence of underground and overhead services
- present and previous site usage
- previous ground investigation data
- location and details of known or suspected contaminants
- other information that can reasonably be obtained from surveys and other sources.

Contractors will use the PCI in preparing their proposal for the work.

Under the CDM Regulations, it is vital to notify the HSE of projects where construction work is likely to

- last more than 30 working days or
- involve more than 500 person days.

Part 3 of the CDM Regulations then applies, in addition to Part 2.

Equipment used in intrusive techniques on contaminated sites, including landfills, must be suitably decontaminated before carrying out maintenance work on the equipment or removing the equipment from the site. The construction phase plan should contain site-specific arrangements for cleaning and decontaminating the plant and identify the measures required to protect site operatives and the environment during this process.

The construction phase plan should list all known or suspected hazards on the site. The desk study or other preliminary investigations (see Section 5.0) should have identified the risks. The construction phase plan must identify and include measures to control the risk to the health and safety of those working on the site, as well as protection of the environment. The risks to those involved in off-site activities, such as the transport or laboratory testing of materials, should also be identified (see Sections 17.0 and 18.0).

The construction phase plan, or its equivalent, should be communicated to all employees on the site and should form the basis of site-specific induction training (see Section 15.0).

Provision of PPE represents the lowest risk control measures in the principles of prevention and protection. Consideration must be given to measures that eliminate or control the risk at the source. Primarily, the consideration of alternative methods of work may achieve this. The construction phase plan, or its equivalent, should contain arrangements for the provision, storage, replacement and disposal of PPE.

The construction phase plan, or its equivalent, should contain a procedure for checking the nature of contaminants and for dealing with the situation where unexpected contaminants are suspected or encountered.

Contractors must not allow work to start or continue unless control measures are in place. For example, intrusive works should not commence until underground services are identified that could be intercepted by the works.

Contractors must, among other things

- check clients are aware of their duties
- satisfy themselves that anyone they employ or engage is competent and adequately resourced
- plan, manage and monitor their work to ensure safety
- ensure that any subcontractor appointed or engaged is informed of the minimum amount of time allowed for them to plan and prepare
- provide workers with any necessary information and site induction (where not provided by a principal contractor) in order to work safely
- comply with welfare requirements and Part 4 of the CDM Regulations
- co-operate with others and coordinate work
- ensure proper workforce consultation on health and safety
- obtain specialist advice where necessary (e.g. when planning high-risk work).

Any contractor, including a subcontractor, must provide information to the principal contractor for inclusion in the construction phase plan.

British Drilling Association
ISBN 978-0-7277-3507-2
https://doi.org/10.1680/sialpic.35072.025

13. Personal protective and site safety equipment

Strict adherence to personal protective equipment (PPE) including respiratory protective equipment (RPE) and site safety equipment requirements is essential. A list of basic PPE and site safety equipment is provided in Appendix C. This should not be seen as exhaustive.

PPE provided for use on sites should include, as a minimum, hard hats, appropriate eye and ear protection, reflective jackets or waistcoats and safety boots or wellingtons with toe and sole protection. The PPE provision may vary in detail depending on the site circumstances and the task being carried out, but will be identified as a risk control measure in the site-specific risk assessment.

Additional guidance on appropriate PPE/RPE on contaminated or potentially contaminated sites is available in the following HSE documents

- dermal exposure (*Managing Skin Exposure Risks at Work* (HSG262))
- gloves (HSG262)
- skincare products (HSG262)
- respirators (*Respiratory Protective Equipment at Work: A Practical Guide* (HSG53)).

RPE should be correctly selected (see HSE guidance HSG53). Appropriate filters and face fitting for individuals are required. Most respiratory equipment manufacturers will advise on suitability, including the use of filters.

It is the responsibility of individual organisations to provide the correct PPE for the work. To ensure the effectiveness of PPE as a control measure, no activity encouraging hand-to-face contact should be allowed, including eating, drinking or smoking while on site without first washing the hands. Implementing suitable working practices reduces the potential ingestion of contaminated solids or liquids.

All parties should wear the proper level of protection before entering the contaminated or potentially contaminated area.

Appropriate impervious gloves should be worn whenever an individual comes into contact with or handles waste (including sharps/punctures), soil, groundwater, samples or other potentially contaminated plant, equipment or materials.

The level of protective clothing should be upgraded if there is any likelihood of external dermal (skin) exposure to unexpected contaminants or substances known to be toxic by the dermal exposure route.

Personnel leaving a contaminated zone and entering a clean zone should observe the site-specific procedure.

On sites where a decontamination unit is required, it should consist of three distinct parts – the first for storing contaminated overalls, footwear and so on; the second providing a high standard of washing facilities, including a shower; and the third allocated for storing clean clothing that personnel will wear on leaving the site. Toilet facilities should be positioned so that washing can occur before use.

Wash-water generated during decontamination of protective clothing and equipment should be transferred to a suitable sealed tank or chamber for appropriate subsequent disposal if found to be hazardous after testing.

Contaminated clothing should only be dealt with by professional cleaners who have been informed of the nature of the possible contamination. Disposable clothing should be disposed of at a designated and approved location off site.

The use of 'off-site' clothing in dirty areas should not be allowed and appropriate site clothing should be provided.

Gas monitoring equipment should be used to detect asphyxiant, toxic or explosive gases on or around the site. Gas monitoring equipment and detectors must be intrinsically safe and relevant under BS EN guidance (see Section 16.6). Equipment should be checked and calibrated following the manufacturers' instructions.

Routine inspections and, where appropriate, calibration of monitoring and safety equipment should always be carried out. The need to upgrade safety equipment should also be kept under review.

British Drilling Association
ISBN 978-0-7277-3507-2
https://doi.org/10.1680/sialpic.35072.027
Emerald Publishing Limited: All rights reserved

14. Health surveillance

Any worker on a contaminated site, or those taking, transporting or testing samples, may require health surveillance. The need for health surveillance should be determined as part of the COSHH assessment. The COSHH assessment will identify risks associated with exposure to fumes, dust, vapours, aerosols and biological agents.

Health surveillance is appropriate where

- a disease or adverse effect may be related to exposure
- this could likely arise in the circumstances of the work
- there are proper techniques for detecting the disease or effect.

Any work with certain chemicals typically needs health surveillance (e.g. cadmium, phenol, arsenic, lead and asbestos). Guidance on working with these chemicals is available (see Sections 20.3 and 20.4).

Further information regarding health surveillance and whether it is required, the process, how to involve employees, medical record format and record keeping can be found on the HSE website (https://www.hse.gov.uk/health-surveillance/).

Occupational monitoring should be routine during intrusive work in order to establish worker exposure to contaminants. It may be appropriate to employ an occupational hygienist for this monitoring.

Details of competence/registration and so on may be obtained from the British Occupational Hygiene Society (BOHS).

British Occupational Hygiene Society (BOHS)
5/6 Melbourne Business Court
Millennium Way
Pride Park
Derby DE24 8LZ
Tel: 01332 298101
Email: admin@bohs.org
Web: www.bohs.org

The BOHS sets standards of competence and holds a register of members who have met education, postgraduate training and experience requirements before passing an entrance exam.

It is recommended that all organisations involved in intrusive activities develop and implement a health surveillance policy. This policy should set down guidelines and procedures for monitoring the health of workers exposed to risk on contaminated sites.

British Drilling Association
ISBN 978-0-7277-3507-2
https://doi.org/10.1680/sialpic.35072.028

15. Site induction

Site inductions can take several different forms. They can be briefing sessions for technical and supervisory staff before they leave the office or drilling crews before they leave the depot. They can also be toolbox talks given on site before work commences and whenever an additional person joins the site team.

The content and scope of the induction is defined by the safety documentation covered in Section 12.0. It is critical that site personnel are aware of the relevant contents of the safety documentation (i.e. the construction phase plan or equivalent) and that there is access to such in an agreed location. Accordingly, the induction should be a formal process that is part of a written procedure. There should be a requirement to record the date, time, attendance and content of the inductions. Those attending should sign a register or some other record form to confirm that they were present and have understood the content of the induction.

The induction's format should be set down in the procedure, with a checklist of the topics to be covered. For site use, pictorial or diagrammatic posters are helpful, enabling the presenter to structure the talk around clearly illustrated subject matter.

Who delivers the site induction will depend on the nature and size of the project, but they should possess sufficient competency to be able to deliver the induction professionally and answer any questions. This is often a supervisor or manager, or the responsibility will be delegated to a sufficiently senior member of the project team.

In any induction, participation in discussion by those attending should be encouraged to promote 'ownership' of safety issues and to heighten awareness. It is only by encouraging interaction that any misunderstandings can become apparent. If there are any misunderstandings, these should be resolved at the induction before work commences.

The induction should be a balanced and open presentation of the issues. Its purpose is to ensure all site personnel have complete awareness and to instigate an appropriate level of safety without overstating the risks or raising issues that are irrelevant to the site.

The site induction should reference the BDA site categorisation concerning potentially contaminated land, the predetermined requirements for PPE and the specific site procedures.

The induction should include general and site-specific issues concerning dealing with contaminants and other potentially hazardous materials that personnel may encounter during work.

The possible content of an induction is provided in Appendix E. This content relates to general safety issues, not just those associated with landfills and contaminated land. The emphasis placed on the section dealing with contamination will depend on the type, severity and extent of the anticipated or suspected contamination.

The induction should also raise awareness of the potential for unexpected hazards personnel may encounter and how such circumstances should be dealt with. Emergency procedures need to be outlined, with a clear indication of responsibilities and the contact details of those who should be immediately informed should an incident occur.

The induction should include increased requirements for personal hygiene and the need to clean hands before and after using the toilet to prevent the transfer of contaminants to sensitive areas of the body. Where a decontamination unit is provided, personnel should be instructed in the correct and safe use of the unit.

Site personnel should be instructed on why smoking or eating is not permitted on contaminated sites, except in designated areas. This instruction and practice are necessary to avoid hand-to-mouth contact because of the risk of ingesting contaminated materials and potential fire or explosive hazards.

The better site staff are informed of the hazards associated with the contaminants they may encounter, the more likely they are to understand the need for the appropriate level of protection and thereby ensure that this protection is in place.

For particularly toxic and/or invasive substances and toxic plants, it is advisable to have data sheets available at the induction, indicating the nature and effects for each contaminant. Several databases offer toxicity information that can be used to obtain the information required to develop these data sheets. Where specific contaminants are identified in the desk study or through ongoing site assessment, the site operatives likely to be affected must be provided with the specific control measures required during the induction. This may require individuals to have received specific training (e.g. plant identification training) where toxic plants may be encountered (e.g. giant hogweed).

Where asbestos has been identified in the RAMS, site staff should – as a minimum – have completed an asbestos awareness training course. If sampling of suspected asbestos is required, a non-licenced work training course appropriate to asbestos in soil should be attended. Further details relating to asbestos in soil can be found in the AGS and CL:AIRE guidance (see Section 20.4).

Hazardous substances can be in solid, liquid, gaseous or microorganism forms. The induction should include reference to each of these to the extent that they are relevant to a particular site. For example, the induction should clearly explain the effects of inhalation together with the potential consequences such as drowsiness, dizziness, headaches, stinging eyes and so on.

The induction should explain the importance of personal gas monitors, the gases or vapours being monitored and the associated trigger levels or ongoing site-wide air monitoring for gases, vapours or asbestos, if these hazards have been identified.

The induction should highlight any contamination risks associated with lone working, or any procedures personnel should follow so that lone working does not occur.

In addition to the induction, site operatives must carry out continual assessment of site conditions to identify if conditions change that may require the revision and amendment of the risk assessment and methodologies (e.g. the issue of disposable gloves to protect from carcinogens absorbed through the skin, such as benzene, may not be all that is required). If the concentrations are sufficiently high to produce fumes that exceed the safe exposure level, breathing apparatus will also be required.

British Drilling Association
ISBN 978-0-7277-3507-2
https://doi.org/10.1680/sialpic.35072.030
Emerald Publishing Limited: All rights reserved

16. Site management

16.1. Site rules

Work site procedures should always be established before work commences so that there is agreement on a safe work system.

The site rules regarding health and safety that should be established on contaminated sites are listed below. However, the list is not exhaustive and there are likely to be specific circumstances at individual sites requiring particular site rules.

- No smoking on site.
- No eating except in designated clean areas.
- All staff to receive a briefing during the site induction detailing the anticipated hazards and safe systems of work to protect them from the particular hazard.
- Appropriate PPE and RPE, where identified as necessary, should always be worn.
- Appropriate welfare facilities and arrangements (in accordance with HSE guidelines and legislation) for personnel should be provided at each work site.
- Lone working is to be avoided.
- No hot works are to be permitted at work sites where flammable gases are suspected.
- Essential comfort breaks must be provided in situations where restrictive PPE/RPE is worn or during periods of very hot or cold weather.

More detailed examples of site rules are now presented.

16.2. Use of PPE, RPE and decontamination equipment

Personnel may not be familiar with using particular types of PPE and RPE, and they must receive training before first use. Many site workers may not be familiar with decontamination facilities and must receive instruction on how they operate under emergency conditions.

Operatives required to wear RPE must have undergone face fit testing, must use the same type of RPE as tested during face fitting and should be clean shaven.

Specific site rules and training will be required to ensure proper operation of clean and dirty areas in decontamination facilities. Properly licensed disposal of used or dirty PPE and RPE may be needed on specific sites.

16.3. Prevention of cross-contamination

Plant and equipment, including sampling apparatus and down-hole equipment, should be cleaned to avoid cross-contamination between sites, work activities and, where appropriate, individual locations or stratigraphic horizons. This cleaning process also minimises the potential of contaminating 'clean' ground, groundwater or the environment and/or compromising technical information obtained from the work.

Methods of cleaning will vary depending on the severity and types of contamination encountered on the site. Cleaning would initially involve the use of tap water and detergent, using a brush if necessary to remove particulate matter and surface films. Steam cleaning or high-pressure water may be necessary with soap/detergent to remove matter that is difficult to remove with a brush. The chosen method must ensure effective cleaning. Therefore, whereas steam cleaning or high-pressure water may be necessary for a Red category site it might also remain a valid option for a Yellow site, depending on the circumstances. However, less stringent methods of cleaning are more likely to be appropriate for a Green category site.

It should be borne in mind that some contaminants such as PFAS (see Section 6.2) may be resistant to cleaning using standard detergents. All detergents should be non-phosphate and all detergent residues should be fully rinsed off before reusing the tools or equipment. Safety data sheets and resulting COSHH assessments should always be reviewed for any materials used in decontamination.

The cleaning area should be well ventilated and personnel should stand upwind if applying solvents. Details of the cleaning, any detergents/solvents used and the location(s) of cleaning should be recorded, kept in the site files and noted in the factual report.

Cleaning should be conducted before leaving the contaminated work site and, where circumstances dictate, between exploratory (or other forms of intrusive activity) hole locations, between contaminated and uncontaminated ground and, in the case of sampling equipment, between samples as necessary or as specified. However, consideration must also be given to maximising resources and reducing the end-disposal cost of potentially contaminated wash-water.

It may be appropriate for cleaning operations to be undertaken at a central location on the work site. This would avoid the need for containing and then disposing of wash-water at each location of intrusive activity. Care needs to be taken to ensure there is no cross-contamination during the moving of apparatus/equipment to the central location. Pads used for washing/cleaning should be bunded and lined with a suitable membrane.

Adequate controls should be in place to handle waste fluids and solids. Wash fluids should be contained and then stored in suitable containers before the issue of geoenvironmental laboratory test results, as these may indicate that the wash fluids will require licensed disposal by a waste disposal specialist. Under no circumstances should wash fluids be released into any watercourse, drainage ditch, other form of surface water or surface water/clean water drains. Before discharge to any public sewer, a permit must be obtained from the relevant utility company or sewerage undertaker.

The storage of waste water/fluids would be in line with the storage of any other chemical or contaminated fluids, including the requirement for secondary containment. If bunds are used, they need to be of a size to hold 110% of the maximum capacity of the largest storage unit.

Consideration must be given to the hazards of close contact with water-borne contaminants generated during the cleaning process, and appropriate additional PPE should be provided to and worn by site personnel (e.g. goggles and nitrile gloves appropriately related to the anticipated contaminants). Wherever possible, efforts must be made to minimise the volume of water generated during the cleaning process. On congested/busy sites, using steam or high-pressure systems will require a designated washing bay to ensure that personnel not involved in the washing are not affected.

Care must be taken to ensure that the operation and servicing of the equipment used for intrusive activities does not cause contamination of the ground or samples by, for example, lubricants, fuels or drilling media. The water used for decontamination must be potable water brought to the site in a pre-cleaned tank if there is no supply available on site. Detergent must be stored and dispensed directly from clean containers.

Groundwater pollution can often be stratified, and vertical (or any other orientation) holes made in the ground provide pathways for contaminant migration. The same applies to ground contamination that may be in one or more horizons only. In this respect, careful consideration needs to be given to the methods used for the intrusive activity, as well as the possible need for the installation of seals between horizons/strata and the use of dedicated samplers (see Section 16.5).

Before work commences, a site-specific plant and equipment decontamination plan should be prepared for all Red and Yellow sites. As a minimum, this plan should include

- identified decontamination area(s)
- items required to clean plant and equipment
- methods for both decontamination and preventing recontamination
- record keeping
- staff health and safety precautions
- waste disposal specific to the contaminants expected.

This plan should be reviewed throughout the work and modified as appropriate if unexpected contaminants are found, the site circumstances change or any aspect of the plan is found to not be providing appropriate protection to personnel or the environment.

16.4. Prevention of contamination by arisings

Certain intrusive activities (e.g. trial pits) can bring significant quantities of hazardous contaminants to the surface. During trial pitting, clean arisings should be stockpiled separately and, if necessary, used as near-surface reinstatement of any exploratory hole. Using polythene sheeting and plywood boards as a temporary measure during site operations can help ensure that there is no contaminated material left at the surface upon completion of the works.

Potentially contaminated groundwater and leachates should be contained within a controlled area, using polythene sheeting and sandbags to prevent seepage into otherwise clean ground. It is essential to prevent the spread of contaminated drill flush and arisings into surface water bodies, including drainage ditches or shallow aquifers.

16.5. Limiting the creation of contamination pathways

A migration pathway is a physical link between contaminated and uncontaminated ground/groundwater. Consideration must be given to the construction design for any temporary or permanent works to avoid the creation of migration pathways. Work should cease if encountered contaminants cannot be controlled during intrusive activities until a revised safe work method is established.

An aquifer protection system can be specified and deployed during the construction of boreholes to prevent the introduction of contaminated soil and/or groundwater into uncontaminated ground and/or groundwater.

A telescopic approach can also be used in the drilling process, using hydrated bentonite seals at contaminated/uncontaminated boundaries or between made ground (fill) and natural ground. This process facilitates casing diameter reduction and mitigates the potential for pathway creation.

Where a monitoring standpipe is required once the borehole has reached its completion depth, the installation should be designed and installed so as to prevent the formation of a preferential migration pathway. The installation should consider pathways for groundwater, leachate ground gas, vapours or contaminants to migrate to or mix with groundwater units of variable quality. Backfilling of the borehole needs to be conducted in accordance with the Environment Agency's *Guidance on the Design and Installation of Groundwater Quality Monitoring Points*. Separate groundwater units shall be sealed from each other, and flow paths shall not be disrupted by pumping or placing materials of lesser permeability. The response zone of the installation should not bridge between contaminated and uncontaminated strata.

16.6. Control of gases and vapour emissions

A range of gases and vapours can pose a hazard to the health and safety of operatives, site personnel and the general public during site construction activities. Common potentially harmful gases and vapours are as follows (please note this list is not exhaustive).

- *Oxygen deficiency* can result from the presence of other gases (e.g. carbon dioxide), the oxidation of metals, burning or geoenvironmental reactions with specific soil types such as glauconitic sands. Oxygen deficiency can result in death by asphyxiation.
- *Oxygen enrichment* can significantly increase the risk of underground fire (e.g. in a landfill or coal measures) or above ground in respect of equipment and materials.
- *Methane* can be generated naturally from coal measures strata, biologically from the decay of organic material, peat and waste, or can originate from manufactured sources (e.g. a gas leak from a fractured supply main). Methane is explosive in the air at concentrations between approximately 5% (lower explosive limit) and 15% (upper explosive limit) by volume (in reality, the range can be 4.4% to 16.5%). Readers should note that methane at concentrations above 15% by volume is also dangerous because there is potential for dilution to bring the concentration down to within the explosive range. At levels above 33% by volume in air, methane is an asphyxiant due to oxygen displacement.
- *Carbon monoxide* is colourless, odourless and highly toxic. It is produced from the incomplete combustion of fuel; in coal measures, the presence of this gas represents an indicator of heating or combustion. Carbon monoxide is toxic at low levels, with concentrations above 400 ppm

leading to discomfort and possible collapse. Higher concentrations increase the rapidity of the onset of health effects and unconsciousness.

- *Carbon dioxide* is toxic and colourless and has a sharp odour and sour taste. At a concentration of 5% by volume in air, breathing will be laboured. At a concentration of 7–10% there is the risk of falling unconscious in the space of a few minutes. It can be encountered in vehicle exhausts, coal measures, peat deposits, landfills (biodegradation of landfill wastes and organic material) and naturally from the geochemical reaction of groundwater within carbonate soils. High concentrations of carbon dioxide can lead to oxygen deficiency: an increase in the concentration will be at the expense of an equal reduction in the concentration of oxygen.
- *Hydrogen sulfide* is a colourless, poisonous gas that can be detected by smell at a concentration of 0.01–0.30 ppm. Strong irritation of the throat and lungs can occur at a concentration of 20–50 ppm. The gas is extremely toxic, with higher concentrations leading to fatigue of the olfactory system. Long-term chronic and severe life-threatening effects occur upwards of 200 ppm. High concentrations of 4.3% to 46% of hydrogen sulfide by volume are flammable and can explode on ignition.
- *Diesel engine emissions* can contain various toxic and/or carcinogenic compounds and particulates.
- *Volatile organic compounds* such as benzene, toluene and xylene are considered carcinogenic and can be highly explosive at concentrations of approximately 1% by volume in air.
- *Hydrocarbon gases* generated by the volatile elements of petrol, diesel and/or oil are flammable at concentrations of between 1% and 7 or 8% by volume in air. Higher concentrations can be diluted to these flammable values.

Where there is a possible risk of gaseous or vapour emissions, the site rules should include regular or constant monitoring during intrusive operations. When using petrol-driven plant where explosive gases or vapours are expected or possible, it is recommended that spark arrestors and/or Chalwyn valves be fitted.

Where necessary and depending on the gas being monitored, intrinsically safe gas monitors such as a four-point gas analyser or a volatile organic gas monitor fitted with alarms should be used to monitor an operative's exposure to hazardous gas or vapour conditions. There may also be the requirement for the use of bespoke monitoring equipment that has been designed to detect specific gases such as hydrogen cyanide, mustard gas and so on. The site rules should include training in using such equipment and adopting procedures if elevated levels of gas or vapour occur.

Any electronic monitoring equipment on the site should comply with BS EN 50104:2019+A1:2023 and/or BS EN 60079-29-1:2016+A1:2022, together with any standards which may be specific for recording a given gas or vapour detection.

Gas levels should be monitored continuously, preferably at the exploratory hole and/or personally with regard to site operatives. Levels should be manually recorded regularly throughout a shift or immediately after encountering any gas or the suspicion of encountering any gas.

Reference should be made to the current short- and long-term exposure limits defined by the HSE in *Workplace Exposure Limits* (EH40/2005).

When defining a work site as a confined space, personnel should be trained to use the equipment that they are provided with to mitigate potential risk from vapours and/or ground gas, which may include RPE. Confined space work is outside the scope of this publication; readers are referred to HSE L101: 2014: *Safe Work in Confined Spaces*.

Where gas is expected, it is essential to organise work sites so that gases are blown away from the work area if possible. The use of forced ventilation can be applied where necessary. This would also apply to activities in buildings or confined spaces where there is a potential for the build-up of gases from the plant and machinery being used. In these circumstances, forced ventilation may be adequate but consideration should be given to conveying the gases using appropriate pipework to a safe exterior location.

The Dangerous Substances and Explosive Atmospheres Regulations 2002 (DSEAR) apply to environments that could potentially have explosive atmospheres. DSEAR sets out requirements for risk assessment, the elimination or reduction of risk, area classification, training and so on. An industry code of practice (ICoP 4) has been prepared on DSEAR by the Environmental Services Association (*Drilling into Landfill Waste*) (see Section 20.6). The Coal Authority's *Guidance on Managing the Risk of Hazardous Gases when Drilling or Piling Near Coal* should also be consulted (Section 20.5). Readers should also consult the HSE documents L138 and INDG370 for further information (Section 20.2).

In some instances, boundary monitoring may be required for the protection of the general public. Gas levels should be monitored continuously, preferably at pre-agreed locations with a competent person (which, depending on the stage of construction, may be identified within the remediation strategy). Levels should be manually recorded regularly throughout a shift or immediately after encountering any gas or the suspicion of encountering any gas. The site rules should include training in using such equipment and adopting procedures should a release occur.

16.7. Control of fugitive emissions

Water suppression should be used wherever possible and controlled using bunding and recirculation/storage surface tanks. An open unlined pit for the storage of water or flush media should not be used. Rotary water and mud/polymer flush should be used as a preference.

The use of air flush is not recommended due to the risk of disease associated with respirable crystalline silica, significantly increasing health risks to site workers and other third parties and damaging the local environment.

16.8. Disposal of excess arisings and other materials

The dispersal of excess arisings on the work site should be avoided and appropriate skips should be available to contain spoil. Excess arisings should be securely contained, transported, tested and then disposed of according to Section 18.0.

Equipment and consumable items purchased, transported and stored on a site are not considered waste if used for the purpose they were purchased or supplied. However, if these materials are discarded or left on site after the completion of works, then they may be considered as waste by the regulators and are therefore subject to the provisions contained within waste legislation. The sustainability of equipment, consumables and other materials with regard to reuse and recycling should be considered.

Contractors must ensure compliance with the appropriate regulatory requirements for classification, recycling or disposal of materials (e.g. https://www.gov.uk/browse/business/waste-environment).

British Drilling Association
ISBN 978-0-7277-3507-2
https://doi.org/10.1680/sialpic.35072.035

17. Handling of potentially contaminated material or equipment

17.1. General

To obtain a representative sample of a bulk material, the principle of sampling may be considered generic no matter which intrusive investigation technique is utilised. However, the methodologies of obtaining a representative sample will vary site by site, as will the handling and testing requirements. Therefore, this section identifies the areas of most significant risk from contamination, from the point of sampling to the reception of samples at a laboratory.

The areas presenting the most significant risks to site personnel and the environment are

- at the point of sampling
- during transportation
- while awaiting or undergoing analysis.

As the intrusive investigation progresses, the sampling and sample handling methodologies should be assessed and reviewed on an ongoing basis to determine if the techniques remain valid. After the definition or revision of the suitability of each sampling methodology, the appropriate PPE should be made available and its use considered either mandatory or worn when applicable (see Note 1 in Appendix C). Further information is available in BS 10175:2011+A2:2017.

17.2. Sampling

At the sampling point, the person carrying out the sampling is likely to be at risk from either the material under examination or the contaminated material on the sampling equipment. Therefore, as a minimum, the sampler should wear appropriate impervious gloves and other PPE, as described in the method statement, to reduce the possibility of dermal contact or transfer of contaminants through the skin. The breakthrough times of gloves related to the contaminants likely to be encountered should be considered.

BS 10175:2011+A2:2017 requires that sampling personnel change their gloves when taking contaminated or potentially contaminated samples.

Where sampling personnel are undertaking liquid sampling, they should use eye protection.

Personnel must use suitable containers to ensure that potentially contaminated material is handled and transported under the appropriate geotechnical or geoenvironmental laboratory methodologies.

17.3. Containers

Suitable containers are required to maintain sample integrity and ensure that the risk of harm by the three primary routes for contaminants to enter the body is reduced to an acceptable level or eliminated for all off-site stakeholders. The appointed geotechnical or geoenvironmental laboratory can provide details on relevant containers and, where appropriate, preservation techniques for the analysis required and associated precautions.

Sample containers should be clean and suitable for the subsequent analysis. Once the sampler has obtained a representative sample, they should fill the appropriate container, ensuring that preservatives are not spilt, lids are sealed tightly shut and the outside of the container is free of all debris. These steps will ensure that subsequent stakeholders are not at risk while handling the containers.

Sample containers should indicate the BDA site categorisation and any other appropriate warnings, such as the presence of 'sharps' or fibrous material, which could be asbestos. Samplers must label the containers distinctly to highlight that they potentially contain contaminated material or that contamination is suspected.

17.4. Transportation

Before transportation, all samples stored on site must be suitably sealed to maintain moisture and they should be kept within a suitable temperature range. These steps are critical for the preservation of geoenvironmental constituents, especially organics. However, samples should be transported to environmental laboratories as soon as possible to ensure the validity of the test procedures and analysis.

All sample containers should be suitably packaged during transportation from the site to the laboratory to enable safe carriage. Samples for geoenvironmental analysis should ideally be double packaged, with each container protected from its neighbour, especially when using glass containers. Personnel may consider double packaging as comprising the sample container and the transportation box. Cool boxes, refrigerated transport and chain of custody provision for specialist geoenvironmental samples must be used.

Samples should be suitably packaged for transportation so that if the vehicle is involved in an accident, the samples will not present a health and safety risk to the driver, any member of the emergency services, the public or the environment.

Some contaminated samples may be considered hazardous goods, such as those that are being tested for radioactive components or explosives, and should be transported under appropriate legislation (i.e. the Carriage of Dangerous Goods and Use of Transportable Pressure Equipment Regulations 2009 implementing the United Nations (UN) Economic Commission's Agreement concerning the international carriage of dangerous goods by road (2017)). Reference should be made to the receiving laboratory to determine transportation requirements.

When the class of a substance is uncertain and it is being carried for further testing, the consigner should assign a tentative class, proper shipping name and UN number based on their current knowledge.

17.5. Laboratory testing

Sampling personnel must inform laboratories of the potential contaminants contained within samples before their dispatch from a site so that they can prepare or review appropriate risk assessments.

It can generally be assumed that analytical/chemical testing laboratories will adopt safe working practices concerning contamination due to the nature of the testing they regularly undertake. Geotechnical testing laboratories must consider whether they can test potentially contaminated samples within the routine laboratory environment or whether they must take additional precautions.

Full details of anticipated or known contaminants should be provided to the laboratory prior to dispatch from the site to ensure that the laboratories and their staff are pre-warned. Some laboratories that have robust procedures in place to enable the safe storage, testing and disposal of samples that may be contaminated may still not be able to accept other specific contaminants such as those potentially containing pathogens or biohazardous materials.

British Drilling Association
ISBN 978-0-7277-3507-2
https://doi.org/10.1680/sialpic.35072.037
Emerald Publishing Limited: All rights reserved

18. Storage and disposal of geotechnical and geoenvironmental samples

Samples taken on Red sites will need to be treated differently to standard samples, both while on site and while at the sample store. Red site samples should be stored separately to other samples taken from Green and Yellow classified sites and/or other zones of the site and should be protected from puncture and crushing, which may inadvertently release the contents. Tubs should be stored upright in protective boxes with sealed lids. Bulk bags should be stored individually and off the ground or kept on protective sheeting to prevent puncture.

Where samples or waste arisings potentially contain hazardous gases or volatile vapours, these should be stored in well-ventilated areas and not in enclosed spaces in a specific clearly demarcated Red sample or waste holding area or store. Further consideration to the risks presented from vapours and gases while the samples are in transport should be made and precautionary measures implemented to ensure that the samples are airtight and the driver of the vehicle is not inadvertently exposed or overcome by potential gases and vapours. These sample types should be transported on open-backed vehicles at all times and in compliance with the Carriage of Dangerous Goods and Use of Transportable Pressure Equipment Regulations 2009, including appropriate COSHH and hazard warning labels on the samples.

It is good practice to inform the persons receiving the samples about the potential risks presented by the substances they contain and how to safely deal with the material in the event of an accidental leak or spillage. This should be communicated to the receiving laboratory and/or the place where the samples are taken to from the site in advance of arrival (i.e. to the sample store or office). A good example is a sample management plan that describes the risk assessment for potentially hazardous substances in the samples, how to safely handle and store the samples along the chain of custody (i.e. site, store and laboratory) and the mitigation and control measures to be adopted in the event of a release.

Whether contaminated or not, personnel should never consider samples of soils taken during a ground investigation to be waste. On completion of any laboratory testing, any sample remaining will be deemed to be waste and must be disposed of appropriately. Waste classification may be necessary prior to disposal to determine the waste type, its carriage and the final waste destination. Advice should be sought from the waste management company who will be responsible for managing the waste to allow the appropriate skip to be provided before the samples are disposed of.

Where site observations or results from geoenvironmental testing on a site classified as Green or Yellow confirm that the samples are contaminated, the risk presented by the materials needs to be reassessed and reclassified if necessary. The current site categorisation for the samples collected (exploratory hole, zone or site) should be reassessed and upgraded. A flow chart summarising the process is presented in Appendix F. Any sample in a store or laboratory relating to the contaminated stratum that has been reclassified as Red should be relabelled with a Red label and stored in accordance with Red site sample requirements. If the samples are in a laboratory and undergoing testing, the laboratory should be notified immediately.

Several UK regulations and European directives control the classification and disposal of waste materials. The EA has provided guidance on how the legislation applies to contaminated soils in *Technical Guidance WM3: Waste Classification – Guidance on the Classification and Assessment of Waste* (Section 20.6).

British Drilling Association
ISBN 978-0-7277-3507-2
https://doi.org/10.1680/sialpic.35072.038

19. Decommissioning and demobilising

At the end of intrusive activities, fieldwork and monitoring, decommissioning and demobilising are required to leave the site safely and ensure that no contaminants are transported off site.

When a site compound has been set up for the contract, all service connections should be severed by a suitably qualified person on completion of the work and left in a safe condition. All site offices, storage containers, fencing, fuel/lubricants storage and site rubbish should be removed from the site.

Upon completion of intrusive works, all plant and equipment in intimate contact with the ground being investigated should be thoroughly cleaned before removal from the site. All hazards introduced as a consequence of the works carried out (e.g. cement, contaminated arisings) should be removed from the site or appropriately made safe.

The plant and equipment clean-up level will depend upon the site's categorisation, as outlined below. Validation of the satisfactory completion of this work may be required. The level of proof should be contract specific.

- *Green sites.* Personnel should wash down vehicles and equipment used on such sites to remove any arisings or debris that could affect the local highway.
- *Yellow sites.* Vehicles and equipment used on such sites should be pressure washed and, if necessary, scrubbed to remove potentially contaminated arisings or debris. The contractor must dedicate a specific area of the site for this activity, and all material must remain stored on site until an appropriate disposal method is determined.
- *Red sites.* Vehicles and equipment on such sites should be pressure washed and scrubbed to remove potentially contaminated arisings or debris. The contractor may be required to validate a satisfactorily completed cleaning process before vehicles and equipment leave the site. This activity should be undertaken in an area of the site specifically dedicated to this work. The materials generated during this process should be stored on site in sealed containers until the appropriate disposal method can be determined.

The specific requirements for decommissioning monitoring installations will depend upon the level and type of contamination present, the ground and groundwater conditions, and the specification requirements. This guide presents example requirements for the decommissioning of boreholes below.

- *Green sites.* Remove secure covers and concrete surrounds, where present, and backfill any installation ducting with inert granular fill or cement–bentonite grout.
- *Yellow sites.* Remove secure covers, concrete surrounds and ducting to 1 m below ground surface. Backfill remaining ducting with cement–bentonite grout and backfill excavations with inert compacted fill; where possible, place cement–bentonite grout using a tremie pipe.
- *Red sites.* Remove secure covers, concrete surrounds and ducting to a depth of 1 m below ground surface, where ground conditions permit. Remove all remaining ducting and associated backfill materials by over-drilling and backfilling with cement–bentonite grout.

Appropriate PPE should be worn for these decommissioning and decontamination activities, including eye protection and a breathing mask if necessary.

British Drilling Association
ISBN 978-0-7277-3507-2
https://doi.org/10.1680/sialpic.35072.039
Emerald Publishing Limited: All rights reserved

20. References and bibliography

20.1. Health, safety and environmental legislation

Borehole Sites and Operations Regulations 1995.

Carriage of Dangerous Goods and Use of Transportable Pressure Equipment Regulations 2009.

Confined Spaces Regulations 1997.

Construction (Design and Management) Regulations 2015.

Contaminated Land (England) Regulations 2012.

Control of Asbestos Regulations 2012.

Control of Substances Hazardous to Health Regulations 2002.

Dangerous Substances and Explosive Atmospheres Regulations 2002.

Health and Safety (Consultation with Employees) Regulations 1996.

Ionising Radiations Regulations 2017.

Management of Health and Safety at Work Regulations 1999; Management of Health and Safety at Work (amendment) Regulations 2006.

Manual Handling Operations Regulations 1992 (as amended) (MHOR).

New Roads and Street Works Act 1991.

Personal Protective Equipment at Work Regulations 1992; Personal Protective Equipment at Work (Amendment) Regulations 2022.

Safe Use of Lifting Equipment: Lifting Operations and Lifting Equipment Regulations 1998 (LOLER).

Safe Use of Work Equipment: Provision and Use of Work Equipment Regulations 1998 (PUWER).

Working at Height Regulations 2005.

20.2. General

Association of Geotechnical and Geoenvironmental Specialists (AGS)

Client's Guide: The Purpose and Use of a Ground Conditions 'Desk Study' (2022).

UK Specification for Ground Investigation. Thomas Telford, London, UK (2022).

British Drilling Association (BDA)

Health and Safety Manual for Land Drilling: A Code of Safe Drilling Practice (2015).

Guidance for the Safe Operation of Cable Percussion Rigs and Equipment (2020).

Guidance for the Safe Operation of Dynamic Sampling Rigs and Equipment (2007).

British Standards Institution (BSI)

BS 22475-2:2011: Geotechnical investigation and testing. Sampling methods and groundwater measurements. Qualification criteria for enterprises and personnel.

BS 22475-3:2011: Geotechnical investigation and testing. Sampling methods and groundwater measurements. Conformity assessment of enterprises and personnel by third party.

BS 10175:2011+A2:2017: Investigation of potentially contaminated sites. Code of practice.

BS 5930:2015+A1:2020: Code of practice for ground investigations.

BS EN ISO 22475-1:2021: Geotechnical investigation and testing. Sampling methods and groundwater measurements. Technical principles for execution.

PAS 128:2022: Underground utility detection.

Building Research Establishment (BRE)

Building on Brownfield Sites: Part 1. Identifying the hazards. Digest GBG 59-1 (2003).

Building on Brownfield Sites: Part 2. Reducing the risks. Digest GBG 59-2 (2003).

Construction Industry Research and Information Association (CIRIA)	Bowman R, Davies P and Baptie P (2019) *What Lies Beneath – Unexploded Ordnance (UXO). Risk Management Guide for Land-based Projects.* Report C785D.
	Guthrie P, Coventry S, Jones M, *et al.* (1999) *Environmental Issues in Construction – A Desk Study.* PR 73.
	Rudland DJ and Jackson SD (2004) *Selection of Remedial Treatments for Contaminated Land. A Guide to Good Practice.* Report C622.
	Rudland DJ, Lancefield RM and Mayell PN (2001) *Contaminated Land Risk Assessment – A Guide to Good Practice.* Report C552.
	Steeds JE, Shepherd E and Barry DL (1996) *A Guide for Safe Working on Contaminated Sites.* Report 132.
	Stone K, Murray A, Cooke S, Foran J and Gooderham L (2009) *Unexploded Ordnance (UXO): A Guide for the Construction Industry.* Report C681.
Environment Agency (EA)	*Guidance on the Design and Installation of Groundwater Quality Monitoring Points.* SC020093 (2006).
	Land Contamination Risk Management (2021).
	Piling into Contaminated Sites (2002).
Health and Safety Executive (HSE)	*Amendments to 'A Guide to the Reporting of Injuries, Diseases and Dangerous Occurrences Regulations 1995'.* L73 (1999).
	Assessing and Managing Risks at Work from Skin Exposure to Chemical Agents. Guidance for Employers and Health and Safety Specialists. HSG205 (2001).
	Controlling Fire and Explosion Risks in the Workplace. A Brief Guide to the Dangerous Substances and Explosive Atmospheres Regulations. INDG370 (2013).
	Cost and Effectiveness of Chemical Protective Gloves for the Workplace. Guidance for Employers and Health and Safety Specialists. HSG206 (2001).
	Dangerous Substances and Explosive Atmospheres. Dangerous Substances and Explosive Atmospheres Regulations 2002. Approved Code of Practice and Guidance. L138, 2nd edn (2013)
	Managing Health and Safety in Construction. Construction (Design and Management) Regulations 2015. Guidance on Regulations. L153 (2015).
	Managing Skin Exposure Risks at Work. HSG262, 2nd edn (2015).
	Reporting Accidents and Incidents at Work. A Brief Guide to the Reporting of Injuries, Diseases and Dangerous Occurrences Regulations 2013 (RIDDOR). INDG453, Revision 1 (2013).
	Respiratory Protective Equipment at Work: A Practical Guide. HSG53, 4th edn (2013).
	Risk Assessment: The Key to Good Health and Safety. INDG163. Revision 4 (2014).
	Safe Work in Confined Spaces. Confined Spaces Regulations 1997. Approved Code of Practice, Regulations and Guidance. L101, 3rd edn (2014).
	The Control of Substances Hazardous to Health Regulations 2002. Approved Code of Practice and Guidance. L5, 6th edn (2013).
	Working with Ionising Radiation. Ionising Radiations Regulations 2017. Approved Code of Practice and Guidance. L121, 2nd edn (2018).
	Workplace Exposure Limits. Containing the List of Workplace Exposure Limits for Use with the Control of Substances Hazardous to Health Regulations 2002 (as amended). EH40/2005 (2020).

20.3. Inorganics/metals

Health and Safety Executive (HSE)	*Arsenic and You: Working with Arsenic. Are You at Risk?* INDG441 (2013).
	Cadmium and You. Working with Cadmium. Are You at Risk? INDG391 (2012).
	Lead and You. INDG305 (2012).

20.4. Asbestos

Association of Geotechnical and Geoenvironmental Specialists (AGS)

Assessment and Control of Asbestos Risk in Soil – Part 1: Protection of Personnel Working on Ground Investigations (2021).

Assessment and Control of Asbestos Risk in Soil – Part 2: Protection of Personnel Working in Geotechnical and Geoenvironmental Laboratories (2021).

Contaminated Land: Applications in Real Environments (CL:AIRE)

CAR-Soil: Control of Asbestos Regulations 2012: Interpretation for Managing and Working with Asbestos in Soil and Construction & Demolition Materials: Industry Guidance (2016).

20.5. Gases

British Standards Institution (BSI)

BS EN 60079-29-1:2016+A1:2022. Explosive atmospheres. Gas detectors. Performance requirements of detectors for flammable gases.

BS EN 50104:2019+A1:2023: Electrical apparatus for the detection and measurement of oxygen. Performance requirements and test methods.

BS EN 61779-3:2000: Electrical apparatus for the detection and measurement of combustible gases. Performance requirements for Group I apparatus indicating up to 100% (v/v) methane in air.

Coal Authority

Guidance on Managing the Risk of Hazardous Gases when Drilling or Piling Near Coal (2019).

20.6. Waste

British Standards Institution (BSI)

BS EN 12457-3:2002: Characterisation of waste. Leaching. Compliance test for leaching of granular waste materials and sludges – Two stage batch test at a liquid to solid ratio of 2 l/kg and 8 l/kg for materials with a high solids content and with a particle size below 4 mm (without or with size reduction.

Environment Agency (EA)

Dispose of Waste to Landfill (2020).

Technical Guidance WM3: Waste Classification – Guidance on the Classification and Assessment of Waste (2015).

Environmental Services Association (ESA)

Drilling into Landfill Waste. Industry Code of Practice. ICoP 4 (2007).

DSEAR Implementation for the Waste Management Industry. Industry Code of Practice. ICoP 1 (2005).

Appendices

British Drilling Association
ISBN 978-0-7277-3507-2
https://doi.org/10.1680/sialpic.35072.045
Emerald Publishing Limited: All rights reserved

Appendix A. BDA site categorisation system

BDA site categorisation	Broad description
Green	Substances that have little potential to cause significant permanent harm to humans. Examples would be uncontaminated natural materials including topsoil, hardcore, bricks, stone, concrete, excavated road materials, glass, ceramics, abrasives, wood, paper, fabrics, cardboard, plastics, metal components, wool, cork, ash, clinker, etc., provided that these do not contain other substances that could be significantly harmful to humans. Note that topsoil and subsoil may be contaminated and that there is a possibility of asbestos or other contaminants being present in otherwise inert areas. In these cases, a Yellow category applies.
Yellow	Substances that are not sufficiently harmful to potentially cause death, injury or impairment of health but nevertheless require protection to be worn to ensure that any health issues do not arise. Examples would be waste food, vegetable matter, household and garden waste, leather, tyres, rubber, latex, electrical goods and fittings, non-toxic metals, bitumen, fuel ash and solidified wastes. Where there is potential for significant volumes of ground gas at concentrations that are toxic, flammable or could cause explosion, then a Red category should be used.
Red	Substances that could subject persons to risk of death, injury or impairment of health. Examples would be any substances that are corrosive, acidic, carcinogenic, cause skin irritation or respiratory problems, affect the nervous system, affect the organs, etc. See above in respect of ground gas or vapours.

Notes

1 A greenfield site would normally be included in the Green category unless there is evidence to indicate a Yellow category.

2 Indiscriminate dumping (or in the case of older sites, uncontrolled/unlicensed landfilling) may have taken place on a site and this should be taken into consideration when determining the appropriate category.

3 Landfill sites or other sites where significant deposits of bound or unbound asbestos occur should have a Red designation. However, other sites may have only very small quantities of asbestos, often present as asbestos cement, which, while presenting a hazard, may not on its own warrant the highest level of protection. In these cases, it may be sufficient to simply add water to the borehole or other form of intrusive activities to prevent asbestos fibres becoming airborne and hence available for inhalation and to wear disposable masks suitable for low levels of asbestos.

4 The presence of radioactive materials on a site is not included in the above guidance and appropriate references on this should be consulted including regulations.

5 A desk study and a risk assessment must be carried out before the site is categorised (see Sections 5, 6 and 7). As part of the desk study, a site walkover must be undertaken. This should look at both the site and the land usage surrounding the site.

6 If the site itself has an indicative categorisation (e.g. Yellow) but it is believed that surrounding site usage may have caused contamination to the site that would be classified as Red, the site categorisation must be upgraded to Red.

7 If, after carrying out the desk study and reviewing the previous usage, there is doubt as to which category should be assigned, the higher of the two categories under consideration should be used.

8 Once the site category has been determined, this should lead to requirements for safe working practices and the use of appropriate PPE and RPE.

9 If, after a category has been assigned to a site, further information becomes available that warrants a change of category then the category should be changed accordingly and all site personnel immediately informed. This includes information from the investigation itself.

10 If changing a site category results in the required PPE and/or safety equipment not being available, the work must be suspended until they are available.

British Drilling Association
ISBN 978-0-7277-3507-2
https://doi.org/10.1680/sialpic.35072.046
Emerald Publishing Limited: All rights reserved

Appendix B. Indicative BDA site categorisation by industry

Former/current site usage	Indicative category
Agriculture (burial of diseased livestock, livestock dip residues)	Green to Yellow
Extractive industries (coal mines, mineral workings, quarries)	Green to Yellow
Energy industries (gasworks, oil refineries, power stations, carbonisation plants, asbestos works and electrical equipment)	Yellow to Red
Metal production (metal processing, heavy industry, electroplating and metal finishing)	Yellow to Red
Non-metal production (mineral processing works, asbestos works)	Green to Yellow
Glass making and ceramics (glass and ceramics manufacturing)	Green to Yellow
Chemicals (oil refineries, petrochemical works, pharmaceuticals, textile and dyes, paint, drum and tank cleaning works)	Yellow to Red
Engineering and manufacturing (heavy engineering, car manufacturing, ship building, electrical and electronic equipment)	Yellow to Red
Food processing (food preparation and animal processing works)	Green to Yellow
Paper, pulp and printing industries (manufacture of pulp, paper and printing equipment)	Green to Yellow
Timber and timber products (wood preservatives and timber treatment works)	Green to Yellow
Textile industries (animal processing works, textile and dyes industry)	Green to Yellow
Rubber industries (chemicals and tyre manufacture)	Green to Yellow
Infrastructure (heavy engineering, docks and railway land, ship building and ship breaking activities, road transport, garages and filling stations, airports)	Yellow to Red
Waste disposal (sewage works, landfill, disposal sites, scrap yards, drum and tank cleaning works)	Yellow to Red
Miscellaneous (high street trades, research laboratories, hospitals, demolition, ordnance)	Green to Red

Notes

1. A careful analysis of the site history will be needed to define potential contaminants, which will then determine the BDA site categorisation scheme.
2. The full range of potential site usages are given in the DoE Industry Profiles and reference should be made to these.
3. Neither these notes nor Appendix A are a substitute for undertaking a desk study or contract-specific risk assessment. Appendix A is designed to form part of the risk assessment process.
4. Use of the table requires that the employer or his representative (which may be the contractor) will have consulted historical maps, records and other documents relating to the site and which may indicate former site usage.
5. For some former site usages, a range of colour categories is shown. This range reflects the potential range of contaminants and the fact that different concentrations of contaminants will affect the categorisation.
6. If no information is available regarding the likely range of contaminants or their concentrations for a particular site use, the highest category shown for that former usage should be adopted.
7. The list of PPE given for a Green site is the minimum level of PPE that is appropriate for work on site. If a site is classified as Yellow, the list of PPE appropriate for the Yellow site must be added to that for a Green site. For a site categorised as Red, the list of PPE for a Red site will be required in addition to that listed for Green and Yellow sites.
8. The list of PPE appropriate to each category of site is a minimum level of PPE. There may be special conditions on a site that require additional PPE or equipment to be used. This can only be ascertained from the contract risk assessment.
9. It should be noted that, during the course of fieldwork, it may be necessary to re-evaluate the site categorisation. It is therefore possible that the site categorisation could be either increased or decreased.
10. Asbestos is a common material on many sites. This should be taken into account during the site assessment.
11. If UXO is suspected or known to be present on a site, advice must be obtained from UXO specialists and work carried out in accordance with CIRIA C681 (*Unexploded Ordnance (UXO): A Guide for the Construction Industry*).

British Drilling Association
ISBN 978-0-7277-3507-2
https://doi.org/10.1680/sialpic.35072.047
Emerald Publishing Limited: All rights reserved

Appendix C. Personal protective equipment and site safety equipment

	Item	Green	Yellow	Red
1	Hard hat	✓	✓	✓
2	Impervious gloves and other forms of hand protection	✓	✓	✓
3	Eye protection (as necessary)	✓	✓	✓
4	Ear protection	✓	✓	✓
5	Overalls	✓	✓	✓
6	Chemical-resistant waterproofs (as necessary)	✓	✓	✓
7	Industrial boots with sole and toe protection	✓	✓	✓
8	High-visibility clothing	✓	✓	✓
9	Fire extinguisher	✓	✓	✓
10	First aid kit	✓	✓	✓
11	Mobile telephone (outside contaminated area)	✓	✓	✓
12	Clean water supply	✓	✓	✓
13	Washing facilities	✓	✓	✓
14	Dust mask[2]		✓	✓
15	Gas mask[2]		✓	✓
16	Disposable overalls (as necessary)		✓	✓
17	Traffic cones and barriers		✓	✓
18	Safety/warning signs		✓	✓
19	Changing/washing facilities		✓	✓
20	Methane detector (as necessary)		✓	✓
21	Carbon dioxide detector (as necessary)		✓	✓
22	Oxygen deficiency detector (as necessary)		✓	✓
23	Other gas detectors (as necessary)		✓	✓
24	Face shield			✓
25	Disposable waterproofs			✓
26	Wellington boots with sole and toe protection			✓
27	Respiratory equipment (as necessary)			✓
28	Decontamination unit			✓
29	Pressure washer/steam cleaner			✓
30	Adequate fencing to prevent contamination spread			✓

Item	Green	Yellow	Red
The following specialist equipment may also be required			
31 Spark arrestors/Chalwyn valves			
32 Air blower (as necessary)			
33 Vertical exhaust stacks (as necessary)			
34 Blow-out preventor			

Notes

1. The use of Items 1 to 34 inclusive needs to be referenced to the contract risk assessment and the BDA site categorisation.
2. The PPE indicated in the table above is the PPE that is recommended to be used on site. The construction phase plan and task-specific RAMS should take precedence.
3. In respect of Yellow and Red sites, Items 14 and 15 must be available on site. Consideration should then be given in each individual case as to whether the contaminants can cause respiratory problems. If they can, then respiratory PPE must be worn, have appropriate filters and be individually face fitted.

British Drilling Association
ISBN 978-0-7277-3507-2
https://doi.org/10.1680/sialpic.35072.049

Appendix D. Flow diagram of key health and safety processes

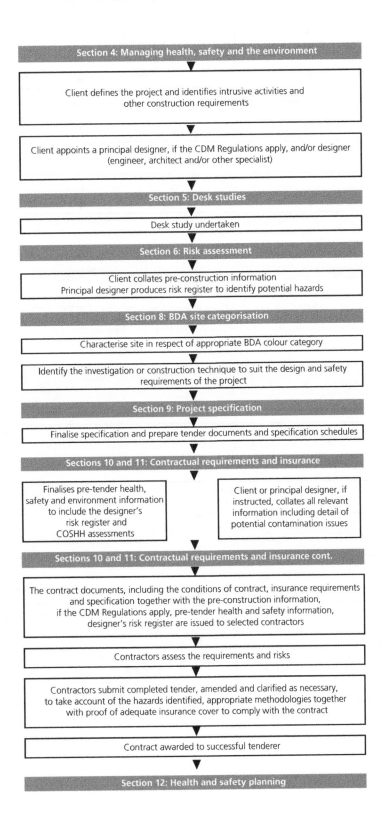

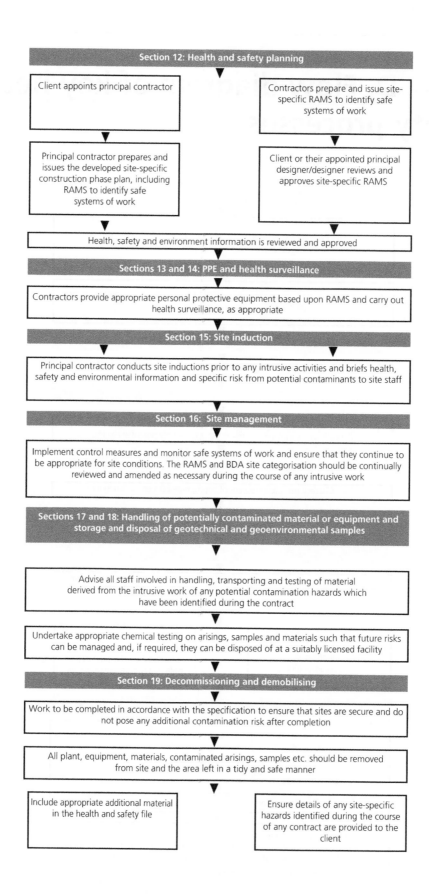

British Drilling Association
ISBN 978-0-7277-3507-2
https://doi.org/10.1680/sialpic.35072.051
Emerald Publishing Limited: All rights reserved

Appendix E. Suggested topics to be covered during site induction

Topic	Details
Project and site details	■ Outline of the project ■ Any specific site rules ■ BDA site categorisation ■ Arrangements for materials and equipment delivery and storage ■ Site tidiness and waste disposal arrangements ■ Welfare facilities – use and cleanliness ■ Parking – reverse parking, neighbours ■ Security arrangements ■ Reporting including accidents, incidents and near misses
Co-operation and liaison with others	■ Other work activities ■ Other specialists involved (e.g. highway traffic management, UXO) ■ Exclusion zones ■ Any designated access/traffic routes
PPE and RPE	■ PPE requirements ■ Special task activities (e.g. noise/dust-creating activities, asbestos) ■ Contaminated materials and COSHH
COSHH	■ Safe refuelling procedure ■ Anticipated/known substances including ground contaminants ■ Sample preservatives
Contaminated land	■ Types of anticipated/known contamination ■ Review of conditions and location ■ Potential hazards, symptoms of exposure and emergency procedures
Safe use of plant and vehicles	■ Good working order – pre-start checks completed daily ■ Ensure protective equipment in place ■ Only to be used by competent persons ■ Reporting of defects
Precautions for services and permit to break ground systems	■ Checking and issue of service plans and PAS 128 surveys ■ Location and/or protection of services ■ Exclusion zones ■ Permit to break ground to be obtained in every instance
Permissions to commence work	■ Instruction from site manager/supervisor ■ Permit requirements – signing permits (e.g. breaking ground, hot works) ■ Check work area/barriers/exclusion zones are in place

Topic	Details
Emergency procedures	<ul><li>Fire precautions – actions, escape route, fire extinguishers, muster points</li><li>Location of first aid kits and confirmation of first aiders</li><li>Location of nearest A&E hospital</li><li>Environmental incident action – spill kits, reporting</li><li>Emergency contacts and contact numbers</li></ul>

British Drilling Association
ISBN 978-0-7277-3507-2
https://doi.org/10.1680/sialpic.35072.053

Appendix F. Process to re-categorise sites

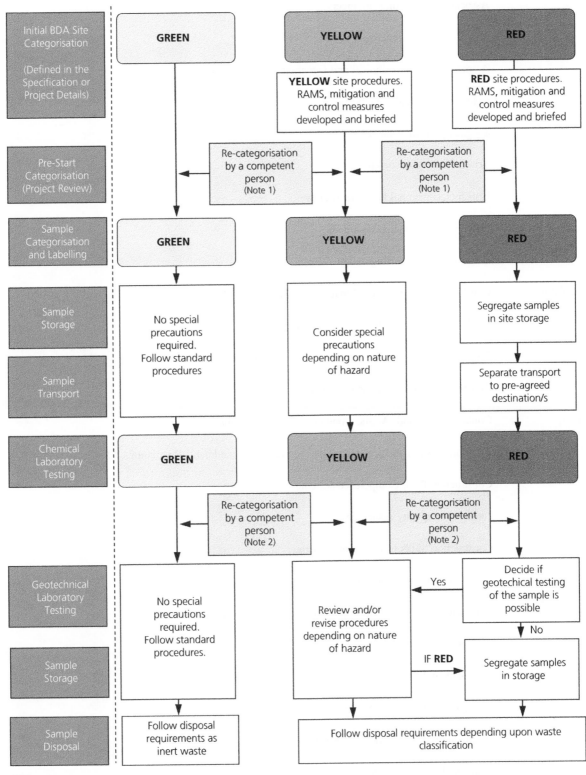

Notes

1. Initial BDA site categorisation typically based on client's/designer's risk register developed from a desk study. Review of categorisation by the contractor is based on desk study and/or site background information (PCI/construction phase plan).
2. If re-classification of the BDA categorisation is required, update RAMS and re-brief personnel.

53

British Drilling Association
ISBN 978-0-7277-3507-2
https://doi.org/10.1680/sialpic.35072.054

Appendix G. Abbreviations

ACE	Associated Consulting Engineers
ACoP	approved code of practice
AGS	Association of Geotechnical and Geoenvironmental Specialists
ALARP	as low as reasonably practicable
APF	assigned protection factor
BDA	British Drilling Association
BOHS	British Occupational Hygiene Society
BRE	Building Research Establishment
CAR	Control of Asbestos Regulations 2012
CDM Regulations	Construction (Design and Management) Regulations 2015
CECA	Civil Engineering Contractors Association
CIRIA	Construction Industry Research and Information Association
CITB	Construction Industry Training Board
CL:AIRE	Contaminated Land: Applications in Real Environments
COSHH	The Control of Substances Hazardous to Health (Amendment) Regulations 2004
CPCS	Construction Plant Competence Scheme
CQA	construction quality assurance
CSCS	Construction Skills Certification Scheme
DoE	Department of Environment
DSEAR	Dangerous Substances and Explosive Atmospheres Regulations 2002
EA	Environment Agency
EPA	Environmental Protection Act 1990
ESA	Environmental Services Association
HASWA	Health and Safety at Work etc. Act 1974
HSE	Health and Safety Executive
ICC	infrastructure conditions of contract

ICE	Institution of Civil Engineers
ICoP	industry code of practice
JIWG	Joint Industry Working Group
LCRM	land contamination risk management
MHSWR	The Management of Health and Safety at Work Regulations 1999
MPQC	Mineral Products Qualifications Council
NEC	New Engineering Contract
NHBC	National House Building Council
NPF	nominal protection factor
NRW	Natural Resources Wales
NVQ	National Vocational Qualification
PCI	pre-construction information
PFAS	perfluorinated and polyfluorinated alkyl substances
PFOA	perfluorooctane acid
PFOS	perfluorooctane sulfonate
PPE	personal protective equipment
ppm	parts per million
RAMS	risk assessment and method statement
RIBA	Royal Institute of British Architects
RIDDOR	Reporting of Injuries, Diseases and Dangerous Occurrence Regulations 2013
RPE	respiratory protective equipment
SCBA	self-contained breathing apparatus
SEPA	Scottish Environmental Protection Agency
SISG	Site Investigation Steering Group
UKATA	UK Asbestos Training Association
UXO	unexploded ordnance

British Drilling Association
ISBN 978-0-7277-3507-2
https://doi.org/10.1680/sialpic.35072.057
Emerald Publishing Limited: All rights reserved

Index

Printed and bound by CPI Group (UK) Ltd, Croydon, CR0 4YY

21/03/2024

14474166-0002